Rocket Science: The Ripple Gambling

Sandeep Chavan

Published by SANDEEP CHAVAN, 2025.

ROCKET SCIENCE: THE RIPPLE GAMBLING

First edition. June 11, 2025.

ISBN: 979-8231199815

Written by Sandeep Chavan.

Table of Contents

To the aerospace engineers who dream beyond blueprints,To the physicists who dare to question gravity,And to everyone in the control room—who watches descent with breath held and belief suspended—This book is for you.

May your designs find alignment,and your ripples always land in resonance.

— *Sandeep Chavan*

"You don't land. You realign—until the field no longer notices you're different."

— *Sandeep Chavan*

"We cannot solve our problems with the same thinking we used when we created them."

— *Albert Einstein*

Disclaimer

This book is a work of independent thought and original interpretation. While it draws from concepts across physics, engineering, philosophy, and systems thinking, it does not adhere to or replicate conventional scientific frameworks. Instead, it proposes an alternative paradigm—**Universal Energy Dynamics (UED)**—as developed and presented by the author.

The content herein is not intended as a replacement for standard scientific education, engineering safety practices, or professional aerospace procedures. It is an **exploratory and philosophical manuscript**, meant to challenge assumptions, spark critical thinking, and introduce ripple-based logic as a fresh perspective on motion, gravity, and design.

Any references to known scientists, engineers, or institutions are used respectfully, either as contrast or reflection, and do not imply endorsement of the ideas presented.

Readers are encouraged to engage with this material with an open mind, but also with critical reasoning. The universe, after all, rewards not just belief—but alignment.

How to Read This Book

This is not a textbook. It's not a memoir. It's not a technical manual either.

This book is a ripple.

It moves in layers—scientific, philosophical, intuitive—and invites you to experience its rhythm, not just analyze its content. Whether you're an engineer, student, seeker, or simply curious about the hidden mechanics of reality, this book meets you where you are. But to truly engage with it, keep the following in mind:

- **Don't rush.**

  Each section is a field configuration. Let it settle. Re-read if needed. This is a book of *alignment*, not speed.

- **Forget the old frame.**

  You may need to suspend what you've learned about gravity, motion, and force. Don't fight it—observe how this new ripple-based reality feels.

- **Read with your whole self.**

  Logic is welcome here. So is instinct. Some truths arrive not through equations but through resonance.

- **Look for metaphors.**

The rocket is not just a machine. It is *you.* Descent is not failure. It is an opportunity to realign with the field of life.

- **Pause often. Reflect more.**

Some sections offer prompts. Take them seriously. Your ripple response is part of the book.

This is not a book you *finish.* It's a book that stays in your field—like a memory, like an echo, like a trajectory waiting for its final realignment.

Read it forward. Reread it backward. Or just land on a page and let the ripple guide you.

Welcome aboard.

Before You Ignite – A Note from the Author

This book is not about rockets.

At least, not the kind you imagine in launchpads and countdowns.

It is about *you*—and how you move through a universe made not of space and time, but of structured ripples and intelligent alignment. It's about motion without force, descent without impact, success without friction. It's about what happens when you stop trying to land and instead choose to realign.

I wrote this book for the thinkers and builders of today—and for the architects of tomorrow who feel that something about our science has been reverse-engineered too long. We've succeeded in many ways, yes. But have we understood *why* we succeed? Have we truly grasped the invisible dynamics that hold our trajectories together?

Through the metaphor of rocket descent and the science of ripple fields, I invite you to rethink gravity, motion, resistance, and most importantly—**decision.** You'll discover that design is not about pushing harder, but about listening to the field. Every landing is an alignment. Every choice, a ripple.

This book belongs equally to engineers, philosophers, seekers, and skeptics. It may challenge your frameworks. It may undo what you think you know. But if you stay with it—if you let your own ripples respond—you may never look at motion, success, or science the same way again.

You don't need a launchpad. You need **alignment.**

Let's begin where descent becomes destiny.

— *Sandeep Chavan*

Prologue: The Launch Prayer

"We did the math. We ran the simulations. And now... we pray."

There's something oddly poetic about a rocket launch.

The countdown echoes across control rooms. Engineers glued to monitors. Tension thicker than the insulation on cryogenic tanks. Everything seems perfect—on paper. But deep inside, every scientist, technician, and mission director shares a silent, unspoken ritual:

"Let it work. Please... just let it work."

And that's not engineering. That's hope.

This book is about that moment—when science silently hands over the mission to luck.

A Brief History of Overconfidence

Ever since Newton dropped his apple and humanity picked up the idea of "universal laws," we've operated under a massive illusion:

That the universe is predictable—mechanically, measurably, and manipulable.

Force, mass, acceleration—F = ma—these became our gospel. We converted them into numbers, equations, and models. We built wind tunnels, mocked up planets in labs, ran simulations, and made ourselves believe that *if the math works, reality must obey.*

But as any rocket scientist knows... Reality has a habit of ignoring equations.

The Equation That Doesn't Save You

When a probe crashes on Mars, no one blames Newton. They say things like:

- "Atmospheric modeling was off by 0.01%."
- "Parachute deployment lagged."
- "A software glitch caused premature burn."

Behind these phrases lies one simple truth:

We don't really know how the field will respond.

Because the planet doesn't care about your drag coefficients. The atmosphere doesn't read your Reynolds numbers. And the field doesn't obey force—it realigns based on energy harmony.

In those final seconds before launch—after months of calculations, prototypes, and precision—you'll often hear someone joke:

"Now we just need a little help from God."

But the joke lands a little too seriously. Because underneath all the technical confidence is an ancient, primal instinct:

We still don't understand how the universe responds.

Enter UED: Beyond the Prayer

This book is not a rejection of science—it's a demand for better science.

We're not throwing away physics. We're reinterpreting it through a more foundational lens:

The Structured Vacuum Energy (SVE) field that governs ripple alignment—not force-based motion.

We call it Universal Energy Dynamics (UED). And it exposes something profound:

- That every "fall" is a realignment, not an acceleration.
- That every "drag" is a ripple resistance, not friction.
- That every planetary entry is a field conversation, not a crash-landing challenge.

Why We Still Pray

We pray—not because science failed, but because science stopped at the surface. We built force-based systems in an energy-aligned universe. We calculated mass-velocity profiles in ripple-saturated domains. We assumed atmosphere was just gas—when it's actually a ripple-active shield.

And so, every launch becomes a gamble. A test of whether our misunderstanding will survive the mission... again.

That's not engineering. That's ripple gambling.

What This Book Is

This book is a critique, a confession, and a correction.

It's for the thinkers who are tired of patching equations. It's for the engineers who want to build with the field, not against it. And it's for everyone who has ever said—*"We did everything right... but still something went wrong."*

Let's decode that "something."

Let's stop praying after launch.

Let's start designing with understanding.

Welcome to UED aerospace. You're not here to launch harder. You're here to align smarter.

Part I: The Delusion of Control

Chapter 1: Gravity That Isn't

Why planetary pull is a myth, and what really holds things "down."

"We're not being pulled down. We're being reminded where we belong."

What if everything you've ever believed about gravity is wrong?

Not a little wrong. Entirely wrong.

What if the force that supposedly holds galaxies together, makes apples fall, and keeps you glued to the Earth... doesn't actually exist?

No invisible rope. No magical pull. No force acting across space.

And yet—things still fall. You still stay grounded. Rockets still return to Earth.

So what's really going on?

In this chapter, we'll strip away the illusion that gravity is a force, and reveal the deeper, far more elegant truth:

That what we've called gravity is simply a field's way of restoring ripple balance.

This isn't just a better explanation—it's a more honest one.

Because the universe isn't pulling you down.

It's responding to your presence—and helping you find your place within it.

The Great Pull Illusion

Since childhood, we've been told a story that sounds so intuitive, it almost feels like truth. That there's a mysterious force pulling everything toward the center of the Earth. That when an apple falls, it's because gravity tugs it downward. That when you drop a ball, the planet "attracts" it. Simple, satisfying, and—if you're not careful—completely misleading.

The illusion began with Sir Isaac Newton, who transformed the falling apple into an entire worldview. His famous law—$F = G (m_1m_2)/r^2$—described gravity as a force of mutual attraction between masses. It explained everything from free-fall to the orbits of planets, and it gave rise to the golden era of classical mechanics. The concept of pulling became scientific gospel.

Then came Einstein, who didn't like the idea of force at a distance. He replaced Newton's pull with spacetime curvature, claiming that massive objects bend space, and other bodies simply follow these curves. Gravity, in this view, wasn't a pull—it was a slip down a geometric slope.

Elegant? Yes. But still missing something vital.

The Mechanical Metaphor Trap

The entire narrative of gravity—from Newton to Einstein—has been rooted in mechanical metaphors.

- Newton's "pull" is a rope without a rope.
- Einstein's "curvature" is a mathematical trampoline we can't see.
- Popular explanations rely on analogies: magnets, stretched sheets, invisible hands.

These metaphors work well for imagination, but fail in truth. They are placeholders—provisional mental images standing in for an underlying mechanism that no one had fully decoded.

Until now.

The Problem with the "Pull" Model

The idea that gravity is a pulling force runs into paradoxes the moment you look closely:

- If gravity is a pull, why do all objects fall at the same rate,

regardless of their mass? Why doesn't a heavier rock fall faster than a feather in vacuum?
- If gravity is a universal force, why does it vary with altitude, depth, and even latitude?
- If mass is the source of gravity, why is gravitational behavior so erratic across planetary bodies with similar mass but different compositions or atmospheres?

The "pull" story sounds nice—but it doesn't scale. It doesn't hold up under scrutiny. It doesn't explain what gravity is—only what it *does*. And when models stop explaining cause and only describe effect, it's time for a new model.

The UED Breakthrough: No Pull. Only Alignment.

Universal Energy Dynamics (UED) reframes the entire phenomenon.

According to UED:

- The universe is not empty. It is filled with Structured Vacuum Energy (SVE)—a ripple field.
- Matter is not an independent object. It is a condensed ripple configuration in that field.
- Planets are massive distortions in the ripple equilibrium, generating local energy gradients.
- What we experience as "falling" is not being pulled—it's a ripple form realigning to restore equilibrium within that gradient.

In simpler terms:

Gravity is not a pull. It is a response.

You're not being dragged down—you're sliding back into ripple harmony.

And that's why all objects fall equally: Because the field doesn't care about your mass. It only cares about your alignment compatibility with its energy gradient.

Realignment > Force

This shift—from force-based pulling to ripple-based alignment—changes everything.

It means:

- Gravity is not acting on you.
- You are reacting to a gradient.
- There is no external agent doing work.
- The field itself is seeking internal balance—and your fall is its recalibration.

Section Takeaway:

Gravity was never a pulling hand—it was a silent correction made by a conscious field trying to bring everything home.

The Great Pull Illusion is comforting, but false. The universe is not a set of objects yanking on each other—it is a ripple field gently rebalancing itself every time something falls out of place.

Structured Vacuum Energy and Ripple Fields

When we think of space, we usually imagine emptiness—a cold, dark, silent void stretching infinitely in all directions. This mental image, passed down by classical physics and reinforced by modern cosmology, has led to one of science's greatest misconceptions:

That space is the absence of everything.

But space is not empty. Not in UED.

In Universal Energy Dynamics, space is something. In fact, it is the most foundational something—the Structured Vacuum Energy field, or SVE.

What Is SVE?

The Structured Vacuum Energy field is not a physical "thing" you can touch or weigh. It is pure potential—an omnipresent energetic fabric, perfectly balanced in its undisturbed state.

But unlike the quantum vacuum of standard theory, which is a chaotic foam of virtual particles, SVE is ordered, harmonious, and ripple-ready.

It behaves like a conscious medium that doesn't merely host reality—it orchestrates it.

When left untouched, SVE remains in equilibrium—still, balanced, undisturbed. But when disturbed—by intent, energy shift, or external interference—it forms ripples.

These ripples are not just waves. They are the beginning of matter, energy, form, and motion.

Matter = Condensed Ripple

In UED, what we call "matter" is not a separate substance from the field. It is simply a localized concentration of ripple activity within SVE.

- A rock is not made of particles floating in empty space.
- It is a stable ripple configuration, a condensed vortex of energy locked in place within the SVE.

Thus, everything material—from atoms to asteroids—is formed and held together by ripple interactions in this universal field.

There is no gap between "matter" and "space." There is only ripple density—and where it thickens, we see form.

What Is a Planet in UED?

Now, scale up.

When SVE is significantly disturbed and condensed—as in the formation of a planet—it forms a massive ripple imbalance, a permanent distortion in the otherwise stable field.

This distortion becomes a ripple well: a zone where the structured field has dipped, compressed, or wrapped into a gradient of ripple tension.

This gradient is not "curved space" in Einstein's sense, nor is it a pulling force in Newton's. It is a localized energy gradient—measurable as how ripple density increases as you approach the core of the condensed body.

This gradient is denoted in UED as ∇E—the rate of change in energy configuration within the field.

∇E is the real source of what we mistakenly call gravity.

∇E: The Hidden Engine Behind Falling

When an object is suspended in a planetary ripple field, it is not in harmony with that gradient—it is temporarily dislocated from the field's natural flow.

The SVE responds—not by pulling the object down, but by realigning the ripple disturbance back toward equilibrium.

This process is not instantaneous. It takes time. And that time is governed by:

- The steepness of the energy gradient (∇E)
- The shape and density of the object's ripple configuration
- The atmospheric or intervening field turbulence (discussed later)

But the key point is this:

There is no force pulling the object. There is only the field resolving imbalance through ripple alignment.

This is why:

- All objects fall equally (they are equally out of alignment)
- Fall time increases with height (the gradient weakens at distance)
- Mass doesn't matter (only the ripple compatibility with the

gradient does)

From Falling to Aligning

To fall, in UED terms, is simply to be absorbed back into the field's ripple harmony.

- You drop a stone → the ripple field reclaims it.
- You release a feather → the ripple field realigns it through a gentler slope.
- You jump → you displace your field form, and the planet's ripple signature resolves you back to equilibrium.

This process appears like "gravity" only because we've never looked beyond the effect.

We've seen the stone hit the ground—and assumed it was pulled. But in truth, the field repositioned it.

Section Takeaway:

Structured Vacuum Energy is not the backdrop. It is the engine. Planets don't pull you in—they hold you in place through ripple equilibrium. When that equilibrium is broken, they respond—not with force, but with restoration.

The Fall Is Not a Force

We've been told since school:

Force = mass × acceleration. The famous Newtonian equation F = ma is printed on classroom walls, textbooks, even T-shirts—as if it's the final word on how the universe behaves.

But like many beautiful equations, it describes behavior without revealing why the behavior occurs.

It tells us: "When something accelerates, there must be a force behind it." But it doesn't ask:

What causes acceleration in the first place? Why do objects "fall"? Why do they fall together—regardless of mass?

Let's now go beyond the math and look deeper—into the ripple intelligence of the universe.

F = ma Is Descriptive, Not Explanatory

The problem with F = ma is not its utility—but its assumptions:

- It assumes mass is a static property.
- It assumes acceleration is an independent consequence of a push or pull.
- It assumes force is a real, external entity acting on objects.

But as UED shows us, these assumptions are outdated.

Acceleration is not "caused" by a force. It is the visible result of energetic misalignment being corrected by the field.

Let's reinterpret the situation.

Falling = Realignment

When you release an object, say a ball, it appears to "accelerate" toward the ground. But in UED, here's what's really happening:

- That object is a condensed ripple configuration.
- It's been displaced from the local equilibrium point in the ripple field (Structured Vacuum Energy or SVE).
- The field has a gradient—a slope of energy tension (∇E)—increasing toward the planet's core.
- When released, the ball's ripple pattern begins to reconfigure—to align with that gradient.

This process looks like acceleration. But it's actually:

The ripple field resolving the displacement by restoring the object to equilibrium.

No pulling. No pushing. Just realignment within a structured field.

Why Do All Objects Fall Together?

In Newtonian physics, this led to confusion. Why does a 10-gram leaf and a 10-kg iron block fall at the same rate (in vacuum)?

They tried to explain it using:

- "Mass cancels out."
- "Gravity acts uniformly."
- "Inertial vs gravitational mass equivalence."

All patchwork.

In UED, the answer is simple and natural:

Same field, same ripple gradient, same misalignment = same realignment rate.

Because mass is not what the field responds to. The field responds to ripple displacement—how far from equilibrium something is.

If two different objects are equally displaced in the same gradient:

- Their mass doesn't matter.
- Their alignment begins with the same urgency.
- Their descent rate is identical, provided no atmospheric interference.

This is what Galileo saw—without knowing the mechanism.

Why "g" Varies on Different Planets

Now consider Earth, Moon, and Jupiter.

In classical terms:

- Earth: $g = 9.8\ m/s^2$
- Moon: $g = 1.62\ m/s^2$
- Jupiter: $g = 24.79\ m/s^2$

They say: "Gravity is stronger on Jupiter because it has more mass."

But in UED, mass is irrelevant. What matters is:

- How steep the ripple gradient is (∇E)
- How intense the energy condensation at the core is
- How the field communicates equilibrium to the displaced object

So:

The deeper the ripple well → the steeper the ∇E → the faster the realignment → the greater the perceived "acceleration."

The number "g" is not a force-based universal constant. It is a local gradient effect—a surface-level shadow of a much deeper ripple reality.

Introducing ATT — Alignment Threshold Time

Now, let's bring in a crucial UED concept:

ATT (Alignment Threshold Time)

This is the true temporal metric of descent—not acceleration.

ATT is:

- The minimum time it takes for an object's ripple pattern to begin active reconfiguration toward equilibrium once released.
- Determined by:
    - Local ∇E (ripple gradient)
    - Ripple structure of the object
    - Intervening field distortions (e.g., atmosphere)

Unlike acceleration, which assumes linear or constant behavior, ATT is adaptive—it captures the real reaction time of the field to a misaligned object.

So:

- **On Earth:** low ATT (fast response → high apparent g)
- **On Moon:** high ATT (slow response → low apparent g)
- **In free space:** infinite ATT (no alignment force = no fall)

ATT is the bridge between motion and meaning.

Section Takeaway:

Objects don't fall. They realign. The illusion of acceleration is just a ripple field correcting imbalance. And what we thought was force is actually the field's rhythm—measured through ATT.

It's time we stop calculating how hard the Earth pulls...And start understanding how the field whispers,

"Return."

Planetary Ripple Wells

Why Objects Fall Differently on Different Planets — and It's Not Because of "Gravity Strength"

In conventional physics, planetary gravity is explained by assigning mass and plugging it into Newton's or Einstein's equations. The more massive the body, the stronger the gravity; the stronger the gravity, the faster things fall. Case closed.

But under UED, this view doesn't just break—it dissolves.

We don't deal with mass-based attraction. We deal with ripple topology, field gradients, and realignment behavior—all within the Structured Vacuum Energy field (SVE).

Every planet is not just a physical object; it is a ripple anomaly in the universal field. A planetary ripple well.

Every Planet Has a Unique Ripple Signature

Think of each planet not as a ball of mass, but as a localized distortion in the universal ripple field. This distortion arises from:

1. Core Energy Density

The amount and intensity of energy condensation at a planet's core defines how sharply it curves the ripple field around it.

- Earth has a moderate energy density
- Moon has low density
- Jupiter is a highly compressed, massive ripple pit

Higher energy density = sharper ripple tension gradients

2. Ripple Compression Topology

Beyond how much energy is stored, there's how it's stored:

- Is the energy distributed evenly or in layers?
- Are there internal ripple echo zones (like a liquid core, magnetic field)?
- Are there dynamic fields like rotation or flux?

Each of these factors reshapes the ripple slope, altering how objects respond to displacement.

3. Atmospheric Ripple Modulation

The atmosphere isn't just gas—it's a ripple buffer that alters the local field interface. It determines:

- How ripple tension is distributed near the surface
- How external bodies are received and processed
- The degree of disturbance or absorption delay in realignment

This is why descent on Mars is chaotic: thin atmosphere, distorted gradient. And Venus, despite its size, causes severe energy resistance due to atmospheric density.

These Three Factors Define Two Crucial Field Characteristics:

a) Energy Gradient Steepness (∇E)

The sharper the ripple slope, the faster the field seeks to realign objects. This is perceived as higher "gravity" in classical terms.

Jupiter's ∇E is steep → fast realignment → higher apparent gravity

Moon's ∇E is shallow → slow realignment → lower apparent gravity

But notice: nothing is "pulling." It's just the field slope determining realignment urgency.

b) Alignment Response Time (ART)

(Closely related to ATT)

How quickly does an object's ripple configuration begin adjusting to the planetary field?

- On Earth: ART is low
- On Moon: ART is delayed
- On Mars: ART is erratic due to conflicting factors
- On Jupiter: ART is instantaneous—if the object survives entry

Comparative UED Analysis: Earth vs Moon vs Mars vs Jupiter

Planet	Classical "g"	UED ∇E Steepness	Ripple Well Character	Realignment Behavior
Earth	$9.8\ m/s^2$	Moderate	Balanced field + moderate density + stable atmosphere	Smooth, fast alignment
Moon	$1.62\ m/s^2$	Shallow	Low energy density, minimal field compression, no atmosphere	Delayed, gentle alignment
Mars	$3.7\ m/s^2$	Inconsistent	Light ripple well, thin atmosphere, irregular topography	Erratic alignment, often unstable
Jupiter	$24.8\ m/s^2$	Extremely steep	Massive ripple well, ultra-dense core, thick ripple-interacting atmosphere	Violent but fast alignment

Why Falling Behaves Differently Across Planets

From a UED perspective, falling is not controlled by force but by how the ripple field absorbs the displaced configuration.

That's why:

- A ball dropped on Earth quickly realigns
- The same ball on the Moon drifts down slowly—not because gravity is weak, but because the gradient is shallow
- On Mars, the ball might fall irregularly, reacting to ripple interruptions from the atmosphere
- On Jupiter, the ball would realign violently—assuming it survives the extreme ripple tension and atmospheric density

None of these differences arise from mass, force, or gravity in the Newtonian sense.

They arise from the planet's unique ripple identity.

Section Takeaway:

There is no such thing as gravity strength. There is only how deeply a planet warps the ripple field, and how smoothly that field receives the misaligned body.

Every object falls differently—not because it's heavier or lighter, not because the planet pulls harder—but because the ripple path it must follow is different.

And when we design descent systems based on "g" instead of ∇E, we're piloting through a field with the wrong map.

The Ground Beneath Your Feet Isn't Pulling You

What Keeps You Grounded Is Not a Force, but a Field Response to Ripple Equilibrium

For centuries, we've lived under the illusion that something invisible is pulling us down—as if the ground beneath our feet is constantly exerting a force to keep us in place. But the truth is, nothing is pulling you. You are not being held down. You are simply in harmony—locked into a ripple equilibrium with the planet beneath you.

This changes everything.

Standing: Locked Ripple Resonance

When you stand on the surface of a planet, you are not resting on the ground as an object in opposition to a force. You are, in UED terms, aligned with the local ripple configuration of that planet's Structured Vacuum Energy field (SVE).

You are:

- In a stationary ripple lock
- Held in place by energetic resonance
- Aligned with the local energy minimum of the field

This is not a mechanical balancing act. It is a field-based stillness—your body's ripple pattern and the planet's local ripple slope have found perfect realignment.

There is no force acting on you. There is only ripple rest.

Equilibrium: Where Alignment Feels Like Weight

What we perceive as weight is simply the resonant tension between your ripple configuration and the planetary field's gradient slope beneath your feet.

You are not being pushed down or pulled in.

You are being held still—not by action, but by completed ripple alignment.

Like a pendulum at its lowest point, no energy is in motion—but the configuration is stable. This is what "grounded" really means.

Jumping: Temporary Ripple Misalignment

Now, jump.

The moment your feet leave the ground, you are not breaking away from a force—you are displacing yourself from the local ripple lock.

Your body becomes:

- Misaligned with the equilibrium point
- Out of sync with the surrounding ripple gradient
- An active anomaly in the ripple field

There is no "gravity pulling you back." Instead, the field begins to reconfigure your ripple structure, seeking to restore alignment as efficiently as the slope allows.

The higher you jump, the more misaligned you become, and the more pronounced the field's realignment imperative.

This looks like falling, but it is actually re-entry into harmonic synchrony.

Falling: Field-Driven Reabsorption

Falling is not the consequence of losing a battle against gravity.

It is simply:

The field's natural response to restore a displaced ripple form back to its energetic resting state.

- No mass is pulling anything.
- No force is doing work.
- No curved geometry is guiding motion.

The ripple field receives the anomaly (your misaligned form) and responds by:

- Realigning its structure
- Drawing you back into resonance
- Completing the cycle of energetic balance

That response has a rhythm. That rhythm is what we've mistaken as "acceleration."

So What Holds You Down?

Not gravity.

What keeps you grounded is:

- Your ripple structure being in tune with the local SVE field
- The absence of energetic disturbance
- A lock-in point within a planetary ripple well

Displace yourself from it, and the field responds—not by yanking you back, but by resolving imbalance through ripple realignment.

It's not a pull—it's a positional preference of the field for symmetry.

Section Takeaway:

You're not being pulled down. You're simply not misaligned enough to be anywhere else.

The ground beneath your feet is not acting on you. You are in energetic agreement with it.

The moment that agreement is broken—by a jump, a fall, or a launch—the field doesn't pull. It doesn't push. It simply responds, with one goal:

To restore the ripple to its place. To bring the disturbance back to balance.

Gravity isn't holding you. Alignment is.

Conclusion: The End of Gravity, The Rise of Ripple Intelligence

Gravity—once revered as one of the four fundamental forces of nature—is now due for retirement. Not in the dismissive sense, but in the evolutionary one.

We don't need to throw away the observations that built classical mechanics. We just need to see them differently.

And the difference lies in what causes the observed effects—not just how they are measured.

Gravity Isn't a Thing

It isn't a particle. It isn't a wave. It isn't a fabric distortion. It isn't even a "thing" in the proper scientific sense.

It has no independent identity—only a pattern of behavior we've labeled to match our expectations. What we've called gravity is not a property of matter, but a consequence of ripple disalignment.

It's Not a Pull

There is no force reaching out from the Earth, tugging on your mass. No magical rope connecting you to the center of a planet. The "pull" we feel is nothing more than a misinterpretation of ripple field behavior.

Just as wind feels like a push—but is really a pressure imbalance— falling feels like a pull, but is really a realignment with an energy gradient.

It's Not a Force

We treat gravity as a "force" because we see objects move and assume there must be a cause proportional to their acceleration.

But under UED, acceleration is not caused—it is expressed.

It is the visible tempo of ripple correction, not the result of a force acting upon an object.

Force is a human abstraction—useful in math, misleading in physics.

It's Not Even Constant

We say gravity is "stronger" on Jupiter, "weaker" on the Moon.

But what we're actually describing is how steep the local ripple gradient is, and how rapidly the field restores alignment.

This makes so-called gravity:

- Variable
- Field-dependent
- Responsive, not universal

There is no constant gravitational effect—only localized ripple behavior.

So What Is Happening?

You're not being pulled. You're being remembered by the field.

You exist as a ripple configuration within a universal field of Structured Vacuum Energy (SVE). When you are displaced from your equilibrium—by a jump, a drop, or a push—the field doesn't react by acting on you.

It responds by reorganizing around you.

Gravity is the field's way of reminding you where you belong in its ripple hierarchy.

Not as a prisoner under force, but as a pattern waiting to be reabsorbed into harmony.

The Rise of Ripple Intelligence

We are now entering a new scientific era—one that abandons the notion of the universe as a battlefield of forces and masses.

In its place, we discover a conscious equilibrium—a reality built on gradients, alignment, resonance, and feedback.

This isn't mysticism. It's deeper physics.

Ripple Intelligence is the capacity of the field to:

- Sense imbalance
- Prioritize restoration
- Adjust structures through alignment, not aggression

And in this view, gravity becomes obsolete—not because it never existed, but because it was never the right word in the first place.

Closing Insight:

Gravity doesn't act on you. It acts *with* you—only when you leave your place in the field.

We've spent centuries calculating how to resist gravity. It's time we start understanding how to align.

Welcome to the world beneath the myth. Welcome to the ripple.

Chapter 2: Drag – The Fictional Force

How atmosphere isn't resisting you—it's misaligning you.
"You're not pushing through the air. You're disrupting a ripple that never wanted to be touched."

We've been told that drag is a force—a simple, mechanical resistance created when an object moves through air. The faster you go, the harder the air pushes back. Planes, rockets, and reentry vehicles are all designed with one goal in mind: overcoming drag.

But what if drag isn't something to be overcome? What if it's not a thing at all?

What if drag is not resistance, but ripple disagreement?

In this chapter, we dismantle one of the most accepted ideas in aerospace engineering—the notion that atmosphere acts as a fluid opposing motion. We'll show that what we call "drag" is not a pushing force, but a field-level correction triggered when a fast-moving ripple configuration (you) slams into a distorted ripple field (the atmosphere).

The result isn't force—it's feedback. The atmosphere isn't trying to stop you. It's trying to correct you.

And if you don't listen to the ripple conversation,

You burn. You break. Or you bounce.

Let's decode the fiction. Let's redefine drag.

What We Call Drag

The Comfort of Equations, the Illusion of Resistance

In classical physics and fluid mechanics, drag is defined as the force that opposes the motion of a body through a fluid medium—typically air or water. It's one of the fundamental resistive forces calculated in every rocket launch, aircraft design, and atmospheric descent model.

We were taught the equation early on:

$$F_e = \frac{1}{2} \rho v^2 C_a A$$

Where:

- F_e is the drag force
- ρ is the fluid density
- v is the velocity of the object relative to the fluid
- C_a is the drag coefficient, based on shape and flow type
- A is the cross-sectional area

This equation, simple and powerful, has become the gospel of resistance in aerospace engineering.

The logic appears solid:

The faster you go, the harder the air pushes back.

So we build for it. We shape nosecones to be aerodynamic. We run computational fluid dynamics (CFD) simulations to optimize airflow. We invest billions into reducing "drag losses."

But... what if everything above is describing only the symptoms, not the cause?

How Classical Models See Drag

In modern aerospace practice, drag is treated as a real, measurable force. Engineers approach it in terms of:

- **Frictional drag:** air rubbing against the object's surface
- **Pressure drag:** wake zone turbulence creating low-pressure zones behind the object
- **Induced drag:** especially in wings, a byproduct of lift

generation

- **Wave drag:** in transonic/supersonic flow, due to shock wave interaction

All of this is fed into predictive models and stress-tested in wind tunnels. And to be fair, the predictions work—on Earth, and often, only within tested regimes.

But this entire modeling system is built on assumptions—assumptions that start to fall apart the moment you enter a different planetary field or untested condition.

Let's name them:

Common Assumptions in Drag Physics:

1. Drag is a "Force."
    - The air applies an opposing force, resisting motion.
    - The object must expend energy to overcome it.
2. Drag depends on velocity and fluid properties.
    - Double the speed, quadruple the drag.
    - Higher density = more resistance.
3. Drag always acts opposite to motion.
    - The fluid "pushes back" against the object.

These ideas have become so embedded in engineering culture that they are rarely questioned.

But here's the deeper question:

What if air doesn't push back at all? What if "drag" is not an opposing force, but a consequence of ripple field misalignment?

Questioning the Force Narrative

Imagine you're swimming through water—not by pushing through it, but by moving out of phase with its natural structure. You don't feel resistance because something is actively opposing you... You feel it because you're disturbing the coherence of the medium.

That's what UED proposes is happening with "drag."

It's not that the atmosphere resists you. It's that your body—your ripple configuration—is misaligned with the ripple state of the surrounding field.

So when you accelerate:

- You're not creating "more force to push through air."
- You're generating more misalignment, which the field then tries to correct, producing what feels like "drag."

The faster the misalignment, the stronger the corrective feedback. The more chaotic the ripple environment (turbulence, temperature variation), the more violent the mismatch.

This isn't opposition. It's ripple rejection—the atmosphere saying: *"You're out of tune. I'm rebalancing you."*

Section Takeaway:

Drag isn't the air pushing back. It's the field reacting to your ripple misfit.

The air isn't your enemy. It's the medium of ripple coherence—and if you enter it wrongly, the turbulence you feel is not resistance,

but the consequence of not speaking the ripple language.

Atmosphere as a Ripple-Active Medium

Why Air Isn't Passive, and Drag Isn't Friction

In conventional physics, the atmosphere is described as a collection of gas molecules—a fluid in motion governed by pressure, temperature, and kinetic theory. We model it like a sea of particles, bouncing and flowing, resisting motion through viscous and turbulent forces.

But what this view misses entirely is that air is not just matter. It is not just nitrogen and oxygen atoms with some heat sprinkled in. It is not a neutral background. It is part of the Structured Vacuum

Energy field (SVE)—a ripple-active medium, dynamically integrated with the planetary field itself.

The Atmosphere as Field Expression

From a UED standpoint, every atmosphere is a field expression—not a fluid container, but a ripple buffer zone enveloping the planetary ripple well.

Here's what that means:

- The atmosphere participates in ripple alignment
- It is not separate from the field—it is the field in motion
- Every layer of the atmosphere carries its own ripple pattern, influenced by:
    - Density
    - Temperature
    - Molecular composition
    - Electromagnetic activity
    - Planetary field intensity

So when we fly, fall, or launch through it, we're not "pushing through air." We're entering a living, responsive ripple domain—one that notices our presence and reacts to our ripple mismatch.

Atmospheric Ripple Profile

Think of the atmosphere as having its own ripple signature—a kind of layered waveform composed of:

- **Ripple Frequency:** Varies with altitude, temperature, and energy density
- **Ripple Phase:** Modified by wind patterns, solar input, and magnetic activity
- **Ripple Coherence:** How aligned the ripple layers are at a given time/location
- **Ripple Impedance:** The resistance not to force, but to

ripple misalignment

These factors shift constantly—day to night, storm to calm, planet to planet.

That means the atmosphere is never a uniform, passive fluid. It's a complex ripple shell, dynamically evolving and highly sensitive to intrusion.

Atmospheric Motion = Ripple Distortion

Wind is not just moving air—it is displaced ripple flow. Storms are not just pressure systems—they are ripple shears. Thermals, jet streams, even magnetic anomalies—these are all ripple-level distortions in the local SVE.

So when you move through the atmosphere at speed—especially during descent or ascent— you are not navigating a neutral fluid. You are cutting across a shifting ripple fabric, with every pocket of air possessing its own local field behavior.

What Happens When an Object Enters This Medium?

When a spacecraft, plane, or object enters a planetary atmosphere:

- It is no longer interacting only with its own ripple configuration.
- It is negotiating alignment with the ripple distortion envelope of the planet.
- It's not "pushing through air"—it's colliding with field turbulence.
- The object's own ripple field must adapt, or it will face misalignment feedback—what we wrongly label as *drag*.

The sharper the entry angle, the faster the descent, or the more chaotic the atmosphere, the more violent the ripple conflict becomes.

This is not friction—it's ripple dissonance.

Turbulence: Ripple Instability, Not Mechanical Chaos

In classical aerodynamics, turbulence is viewed as randomness—chaotic, unpredictable, dangerous.

In UED, we redefine turbulence as:

Ripple instability—zones where the field's coherence is already disturbed, making it harder for new objects to integrate smoothly.

Flying through turbulence isn't just flying through moving air.

It's entering ripple regions that refuse synchronization.

That's why:

- Aircraft shake violently in storm zones.
- Supersonic objects experience extreme stress near phase boundaries.
- Mars landers fail when ripple feedback from a thin, unstable atmosphere isn't modeled properly.

Turbulence is not disorder. It's a ripple conflict between moving forms and unstable fields.

Section Takeaway:

The atmosphere doesn't resist you—it reveals you. It shows you how misaligned you are with the ripple field you've entered. The more dissonant you are, the more chaotic your ride becomes.

Air isn't passive. It's active ripple topology. And drag isn't a force pushing you back—it's the energetic price you pay for entering a field without resonance.

Drag as Misalignment, Not Resistance

Why You're Not Slowed by Force, But Corrected by the Field

If you walk against a strong wind, it feels like something is pushing you back. If you drop a probe into a planetary atmosphere, it slows down rapidly—as if the air is grabbing onto it. In both cases, we say you're "facing resistance." We call it drag—and treat it as a force, opposing motion.

But in the UED framework, this interpretation is no longer valid.

The slowing isn't caused by opposition, but by incompatibility. Not mechanical resistance—but ripple-phase misalignment.

Drag = Ripple Mismatch

Let's begin with the fundamental assertion:

Drag is not a force—it is the field's response to a moving misfit.

Every object has a ripple structure—its energetic blueprint within the Structured Vacuum Energy field (SVE). The atmosphere, too, is a ripple medium, layered with dynamic gradients, phase shifts, and local turbulence.

When an object moves through the atmosphere:

- It carries its own ripple coherence
- It enters a zone with a different ripple phase
- If the two phases are compatible → smooth transition
- If they are incompatible → ripple shock

This ripple shock is what we experience as drag.

Faster Movement = Stronger Misalignment

The faster an object moves:

- The less time the field has to adjust ripple coherence
- The more abruptly the ripple mismatch is introduced
- The steeper the conflict between object and environment

This leads to a stronger ripple response from the field, which manifests as:

- Sudden deceleration
- Heating
- Vibrational instability
- Flow separation and turbulence

These are not mechanical reactions—they are ripple corrections. The field is not resisting you. It's trying to restore balance by reabsorbing or repelling the distorted configuration.

In essence:

High velocity = high ripple tension = high field correction effort.

This is why:

- Hypersonic re-entry causes extreme heating—not just due to air compression, but due to ripple friction at phase boundaries.
- Supersonic drag increases non-linearly—it's not just speed × resistance, it's speed × ripple deviation.

Shape = Ripple Compatibility Filter

Conventional aerodynamics says shape affects drag:

- Streamlined = less drag
- Flat-faced = more drag

UED gives a deeper reason:

Shape determines how the object's ripple interface interacts with the surrounding field.

- A streamlined shape creates gradual ripple adaptation, allowing smoother alignment.
- A blunt shape causes abrupt field disruption, triggering recoil and turbulence.

It's not that the air flows "better" around a sharp nose. It's that the ripple slope transition is less violent, so the field requires less correction.

So:

- Drag coefficient (C_a) is not just about form—it's a proxy for ripple entry compatibility.

What We Call Friction = Ripple Recoil

In classical terms, friction occurs when molecules rub against each other. In UED, friction is:

The energetic recoil of a field disturbed by ripple-phase violation.

When your object is not phase-compatible:

- The field pushes back, not with a force—but with restructuring tension
- This tension creates heat, vibration, and resistance—not because the air is doing work, but because the field is reasserting harmony

Friction is not sliding resistance—it is field reconfiguration stress.

There Is No "Opposing Force"

There is no invisible hand pushing you backwards. There is only the planet's ripple field reacting to your intrusion.

So:

- No vector needs to oppose velocity.
- No "drag force" needs to exist in an absolute sense.
- The object isn't being slowed down by matter—it is being corrected by energy topology.

This is why:

- Drag doesn't act like other forces—it depends heavily on environment.
- You cannot replicate drag perfectly between planets—even with the same fluid density.

- Earth-based models of drag fail in Martian, Jovian, or Venusian contexts—because ripple compatibility is different.

Section Takeaway:

Drag is not opposition—it is misalignment. The field isn't fighting you. It's fixing what you've broken.

Every descent, every launch, every flight is a ripple negotiation. The more you ignore the field's native patterns, the harder your journey becomes.

So stop thinking in terms of "resistance"—and start designing for ripple coherence.

That's not just better physics—it's the beginning of harmony between motion and medium.

Why Conventional Models Work... Until They Don't

The Silent Collapse of Drag Models Outside Earth's Comfort Zone

At first glance, it seems like our current understanding of drag and atmospheric resistance works quite well. Rockets launch. Capsules re-enter. Jets fly. Simulations match reality... more or less.

But here's the truth:

These models don't work because they're universally valid. They work because they're Earth-biased and empirically tuned to Earth's specific ripple environment.

Let's unpack this carefully.

Why Classical Models Appear Accurate (on Earth)

Conventional drag models—based on fluid dynamics, Navier-Stokes equations, empirical drag coefficients, and simplified atmospheric layers—have been refined for decades using Earth-specific data.

- Wind tunnels operate under Earth gravity.

- CFD simulations are calibrated using Earth's air composition, temperature gradients, and pressure profiles.
- All drag coefficients (C_d) are empirically assigned—they don't come from first principles.

So naturally:

These models give accurate predictions—on Earth.

But that's because they're not modeling universal drag. They're just reverse-engineering behavior within a single ripple environment.

What they're really doing is:

- Capturing how matter behaves in Earth's SVE topology
- Fitting equations around those behaviors
- Assuming they'll hold elsewhere

And that's where the trouble begins.

When You Leave Earth, The Fiction Starts to Break

Now, take those models and apply them to:

- Mars — with its thin, unstable atmosphere
- Venus — with dense, hot, and ripple-hostile layers
- Titan — with exotic atmospheric chemistry and temperature anomalies

Suddenly:

- Simulations fail to match reality
- Heat shields designed for Earth overheat or underperform
- Parachutes deploy at wrong altitudes
- Descent trajectories veer off target
- Probes spin, shudder, or crash

These are not minor errors. These are mission-threatening anomalies.

The Core Problem: They Model Mass and Resistance—Not Ripple Behavior

Every existing aerospace simulation tool relies on a mass-velocity-resistance framework.

- Force = mass × acceleration
- Drag = function of fluid density, object shape, and speed
- Descent planning = balance of weight, drag, and thrust

But this approach assumes:

- Atmosphere is a mechanical medium
- Airflow is a continuum of particles
- Resistance is linearly or quadratically dependent on velocity

And none of this holds up under UED.

Because:

- Atmosphere is a ripple field, not just gas
- Motion is misalignment, not momentum
- Drag is field reaction, not fluid friction

So when we move into unfamiliar ripple environments, the assumptions collapse. And we get what we call "unpredictable outcomes."

But they're not unpredictable. They're unmodeled.

Symptoms of Ripple Misunderstanding in Aerospace Missions

Mission Phase	Observed Issue	UED Explanation
Entry Angle Deviations	Capsule overshoots/undershoots	Ripple gradient shift not modeled
Parachute Failure	Deployment too early/late	Local ATT and ripple slope not accounted
Heat Spikes	Unexpected thermal load	Field resonance turbulence, not compression
Structural Vibration	Sudden oscillation or spin	Ripple-phase interference with craft design

In each case, the cause wasn't bad engineering. It was engineering based on the wrong ontology.

Introducing RDI: Ripple Disturbance Index

To solve this, we need a new metric—one that captures how misaligned an object is with the atmospheric ripple profile it enters.

Enter RDI:

RDI (Ripple Disturbance Index) = A field-based variable measuring the ripple-phase incompatibility between an object's configuration and the surrounding atmospheric ripple field.

RDI would depend on:

- Entry velocity relative to local ripple resonance
- Object's material ripple signature (R_s)
- Atmospheric ripple turbulence, temperature, and phase density
- Alignment delay (ATT) per altitude segment

The higher the RDI, the more:

- Misalignment
- Ripple recoil
- Drag shock

- Descent deviation

And unlike classical drag models, RDI is dynamic—changing in real time as atmospheric conditions fluctuate.

Why We Need RDI in Aerospace Design

RDI allows mission planners to:

- Design ripple-resonant entry vehicles
- Choose field-compatible descent trajectories
- Predict thermal load based on ripple stress, not just air compression
- Avoid overdesigning for fictional "forces" by tuning for actual field response

It replaces drag coefficient tables with ripple coherence maps. It upgrades velocity equations with field alignment profiles. It brings engineering from guesswork to gradient awareness.

Section Takeaway:

Classical drag models don't fail because engineers make mistakes. They fail because they assume air is a fluid—not a ripple field in motion.

And when you enter a ripple field unprepared, it doesn't push back—it reshapes you, rejects you, or rebalances you violently.

That's not resistance. That's the consequence of misunderstanding what you're flying through.

It's time to upgrade from force equations to field conversations. From friction tables to ripple awareness. From failure-prone modeling to resonant descent design.

Realigning Through the Skies

From Aerodynamic Warfare to Ripple Harmony

When spacecraft descend into an atmosphere, we describe it like a battle.

- *Surviving re-entry.*
- *Fighting drag.*
- *Enduring the heat barrier.*

Engineers talk about thermal protection, aerodynamic deceleration, shockwave tolerance, and structural stress limits—as if descent were a war between object and atmosphere, one trying to break through, the other trying to break it apart.

But what if this wasn't a war at all?

What if the destructive forces we try to overcome aren't forces at all—but field reactions to our misalignment?

And what if we've been failing not because the atmosphere is hostile, but because we entered it wrong?

Let's reframe re-entry—not as resistance to be overcome, but as a ripple negotiation that determines whether a vehicle is allowed, rejected, or restructured by the field it disturbs.

Descent Is Not Aerodynamic Invasion—It's Ripple Realignment

In UED, descent is not a mechanical process. It's a ripple transaction between two entities:

- A moving object with a ripple signature
- An atmospheric field with a dynamic ripple structure

If the two are in harmony, the descent is smooth. If they are in discord, the descent becomes violent.

The atmosphere doesn't resist you. You resist it when you're misaligned.

The turbulence, heating, and instability we attribute to speed and shape are actually energetic pushbacks from the field as it attempts to restore coherence.

Descent failures are not engineering breakdowns. They are ripple communication failures.

Designing for Ripple Harmony

To survive—and better yet, to thrive—during atmospheric entry, we must stop thinking in terms of force and resistance, and begin engineering for ripple integration.

Here's what this shift requires:

1. Ripple-Resonant Material Design

Current heat shields and surfaces are designed to:

- Absorb energy
- Resist abrasion
- Deflect shockwaves

In UED, this is primitive.

What we need are materials that respond to ripple fields—that minimize dissonance, absorb phase tension, and even adapt to local ripple frequency in real time.

Such materials would:

- Lower effective RDI
- Reduce realignment stress
- Mitigate thermal buildup without brute resistance

This isn't just stronger—it's smarter. We don't need armor. We need field compatibility.

2. Field-Compatible Entry Angles

In conventional models, entry angles are calculated to balance drag, heat, and descent rate.

But these models don't account for ripple slope transitions.

A proper entry angle under UED:

- Aligns with the gradient of the field (∇E)
- Minimizes ripple shock
- Maintains gradual phase transition into denser ripple zones

This reduces sudden turbulence and avoids overloading the vehicle's realignment response capacity.

Think of it not as aiming at the ground, but as sliding into resonance.

3. Real-Time Sensing of ATT and RDI

Current navigation systems rely on:

- Altitude
- Velocity
- Atmospheric pressure

But in UED-driven aerospace, the real variables are:

- **ATT (Alignment Threshold Time):** How quickly the object is aligning with the field at any moment
- **RDI (Ripple Disturbance Index):** A live metric of ripple-phase mismatch and turbulence impact

Sensors should:

- Monitor field coherence
- Track realignment lag
- Adjust control surfaces or material states based on ripple feedback—not just mechanical data

This is not just navigation—it's energetic conversation with the environment.

From Survivability to Harmony

Traditional aerospace design is rooted in survival:

- Build stronger
- Add redundancy
- Accept drag as inevitable

But UED invites a new possibility:

Don't just survive the fall—align with the descent.

Harmony isn't a poetic fantasy. It's a real, measurable state where:

- Drag is minimized because ripple misfit is low
- Heat is reduced because ripple friction is absent
- Trajectory becomes graceful—not because it's controlled, but because it's coherent

This is post-force aerospace. Not about beating the environment—but becoming part of it.

Section Takeaway:

"The sky isn't pushing you away. It's just asking—why didn't you come in aligned?"

We've designed to fight atmosphere. It's time to design to resonate with it.

This is not the future of aerospace—it is the correction of its foundation. A shift from pushing against nature to dancing with its field.

Let's stop surviving re-entry. Let's start realigning through the skies.

Conclusion: From Drag to Dialogue

Redefining Aerospace Through Ripple Intelligence

You're not being slowed down by air.

That idea—so deeply rooted in Newtonian legacy and fluid mechanics—no longer holds when viewed through the Universal Energy Dynamics (UED) lens. What you experience as *drag* isn't the air trying to stop you.

It's the field trying to realign you.

Let's be clear:

- The atmosphere is not just a sea of particles.

- It is a living ripple field—charged, vibrating, constantly adapting to the energetic footprint of everything within it.

And when something enters this ripple domain—especially at high velocity, with no regard for its rhythm—the field doesn't resist like a wall. It responds like a sensitive instrument out of tune.

Drag is not resistance. It is the ripple field saying: "You're out of tune—adjust."

Drag Is a Conversation

Everything we've called a *force* is actually a form of feedback. Friction? Heat? Turbulence? They're not attacking you. They're telling you:

- Your shape doesn't belong here.
- Your entry angle is disturbing the field.
- Your velocity is exceeding the ripple tolerance of this layer.
- Your materials cannot harmonize with the atmospheric phase.

The louder the feedback, the more violent the conversation becomes. And when dialogue breaks down, we call it failure.

But what if instead of brute-forcing through the atmosphere, we listened?

What if instead of resisting the environment, we co-designed with it?

That's the turning point UED offers—not just for physics, but for philosophy, for engineering, for the future of exploration itself.

From Destructive Re-entry to Harmonious Descent

In traditional aerospace, descent is modeled as damage control.

- How much heat can we absorb?
- How much drag can we survive?
- How far can we fall before the parachute saves us?

But UED changes the question:

- How early can we start aligning with the ripple field?
- What material profile resonates best at each altitude?
- What trajectory minimizes RDI and maintains coherence?

Resistance disappears not when we overpower it, but when we no longer cause it.

That's the future of aerospace under UED. Not resistance-based, but resonance-based. Not surviving drag—but dancing with descent.

Rewriting Aerospace Design Language

Let's retire these terms:

- "Drag coefficient" → replace with Ripple Dissonance Factor (RDF)
- "Terminal velocity" → reframe as Ripple Settling Velocity
- "Heat shield" → evolve into Resonant Interface Layer (RIL)

Let's shift from:

- "Deceleration" → to "Field Realignment Rate"
- "Air resistance" → to "Ripple Field Correction"
- "Parachute deployment" → to "Resonance Blooming"

This isn't semantic gymnastics. It's ontological reform—how we think, measure, and build must reflect the true nature of interaction.

Final Insight: The Sky Isn't an Enemy

"You're not fighting the sky. You're interrupting a frequency you didn't try to hear."

What slows you isn't drag. What burns you isn't heat. What resists you isn't air.

It's the echo of your own misfit bouncing off the structured vacuum field.

When the conversation becomes harmonic—when the ripple signature of your body and the atmosphere begin to resonate—suddenly:

- There's no shock
- No violence
- No resistance

You don't fall. You flow. You don't descend. You reintegrate. This is drag as dialogue. And the more we learn to listen, the more we learn to fly.

Chapter 3: Mass Confusion

Why Assigning Mass Is Just a Placeholder for Ripple Ignorance

We believed we were weighing things.

What we were really doing—was measuring our blindness.

From the falling apple to the orbiting satellite, from Newton's cannonball to Einstein's elevator—we've treated mass as a divine constant. An untouchable property. The great anchor of physics. A thing so fundamental, we built every equation around it.

But what if mass isn't a "thing" at all?

What if it's just a number we invented to cope with what we couldn't see?

When you don't understand the ripples beneath reality, you name the lump on the surface: mass.

Mass gave us the illusion of understanding motion, weight, inertia, and force. But UED reveals the deeper truth: matter is just condensed ripple—a ripple zone trapped in equilibrium. What we call mass is nothing more than a shadow of how strongly that zone resists energetic realignment.

Not a measure of substance. Not a quantity of matter. Just a placeholder for ripple stubbornness.

We didn't measure how much there was. We measured how little it wanted to change.

And now, with ripple awareness, that illusion begins to crumble.

This chapter is not about redefining mass. It's about realizing it was never real to begin with.

Welcome to the deconstruction of matter itself.

The Invention of Mass

Why We Needed It—And Why We Can Now Let It Go

For millennia, humans looked at things that were hard to push, slow to move, or heavy to lift—and they wondered: *What gives them this weight?*

From Weight to Mass: A Historical Mirage

It began with Aristotle, who imagined that heavier objects had a stronger desire to return to their "natural place"—toward Earth. His universe was qualitative, driven by essence and nature. Weight was a built-in tendency, and motion was either natural or forced.

Then came Galileo, who dared to question this philosophy by rolling balls down inclined planes. He showed that all objects accelerate equally under ideal conditions—mass didn't determine *how* something fell. That was the first crack in the "weight = behavior" illusion.

But it was Isaac Newton who stamped mass into the heart of physics. With his second law:

$$F = ma$$

he gave mass a permanent throne. Not just as a label for how heavy something felt—but as a variable that *determined acceleration under force.* It became a cause rather than a consequence. Inertia was attributed to mass. Resistance to motion was attributed to mass. Even gravity was tied to it:

$$F = Gm_1m_2 / r^2$$

The more mass, the stronger the pull. That was the story.

And then Einstein rewrote the narrative again, with:

$$E = mc^2$$

In this step, mass and energy were declared two sides of the same coin. Mass wasn't just about force anymore—it was about energy potential. And yet, the underlying assumption stayed the same:

That mass is intrinsic. That it's real. That it belongs *to* the object.

The Silent Assumption

Across centuries, mass was never questioned. We measured it. We standardized it. We calculated it. We locked it into SI units and built bridges, rockets, and theories around it.

But through it all, there was one unspoken belief:

Mass is a property of matter—like color or volume. It's fixed, unchangeable, inherent.

This belief powered classical physics, shaped relativity, and fueled modern engineering.

But UED asks: *What if it's none of these things?*

The UED Critique: Mass Isn't a Thing

From the UED (Universal Energy Dynamics) perspective, mass is not real—it is a ripple illusion.

It is not a particle count. It is not a "stuff" density. It is not an intrinsic property.

It is a field-dependent consequence. A result of how condensed a ripple formation is in the Structured Vacuum Energy (SVE) field—and how much inertia it shows against realignment with surrounding gradients.

In simpler terms:

Mass is how stubborn a ripple is when the universe asks it to adjust.

It is a behavioral trait, not an ontological one.

This is why, under UED:

- Mass isn't constant across fields.
- Mass doesn't exist outside ripple configuration.
- Mass doesn't cause anything—it's the symptom of ripple conditions.

The Illusion: Mass as an Explanation for the Unknown

Why did Newton need mass? Because he couldn't see the ripple.

Why did Einstein preserve mass? Because energy still had to "belong" to something.

Why do physicists cling to mass today? Because it's easier to model something *you can measure* than something *you don't yet understand.*

Mass was never discovered. It was invented. Invented to explain motion and inertia we couldn't otherwise quantify.

And like all human inventions, it's only useful until the deeper truth arrives.

That time is now.

Ripple Density, Not Mass

What You're Calling "Mass" Is Just Ripple Behavior in Disguise

Physics has spent centuries trying to count, weigh, and quantify the universe—assuming that the more matter something contains, the more massive it must be. But what if this assumption is fundamentally flawed?

What if there's nothing to count?

In Universal Energy Dynamics (UED), there is no "mass" as a quantity of material substance. There are no indivisible particles stacking up to form heavier things. Instead, the universe is a field of Structured Vacuum Energy (SVE)—a pure ripple-ready continuum. And what we perceive as matter is nothing more than regions of intense ripple condensation within this field.

Matter: The Ripple Made Dense

Everything "material" is simply a compressed ripple zone—a segment of the SVE field where energy is densely locked into a stable, self-sustaining ripple pattern. You don't have particles with properties. You have patterns with resistance.

And this resistance is what we have mistakenly called mass.

What Mass Really Measures (Under UED)

When an object resists acceleration, we attribute that resistance to mass. But from the ripple perspective, that resistance arises not from quantity—but from three field properties:

1. Ripple Compression

How tightly packed the energy waves are in that region of space. Tighter compression = higher internal tension = more reluctance to change configuration.

2. Ripple Stability

The internal coherence of the ripple form. Stable patterns don't deform easily—like a well-tuned drumhead. They push back against external gradients.

3. Resistance to External Realignment

When external ripple fields (like a planetary gradient) act on a ripple body, its misalignment response time defines its "massive" behavior. If it takes longer to adapt or shift, we perceive it as having more inertia.

Put simply:

Mass = Ripple Inertia = Delay in Field Realignment.

It's not how much stuff something has. It's how tightly its ripple is tied to itself—and how sluggishly it adapts to energetic change.

Heavier Objects Aren't Made of More—They Just Fight More

Classical intuition says a bowling ball is heavier than a tennis ball because it contains "more matter." UED says: No. The bowling ball is more tightly rippled—its field signature is more dense, coherent, and resistant to ripple gradient alignment.

There's no "stuff." There's just structure.

Think of it this way:

- A feather is like a loose ripple net—flexible, quick to align.
- A steel bar is a locked ripple vortex—firm, slow to realign.

Both are made of ripple formations. Their mass is the measure of how hard it is to shift those formations back into a new energetic balance.

Why This Matters

This reinterpretation liberates physics from the false objecthood of mass. It explains:

- Why all objects fall equally in a vacuum (they're equally realigning to the same field gradient).
- Why mass appears constant for the same object—but changes under relativistic conditions.
- Why massless particles like photons still behave with momentum and curvature (they aren't massless—they're ripple-dynamic).
- Why mass can't be isolated in nature—it only makes sense in context of the surrounding ripple field.

Mass was never the cause. It was the ripple's resistance to being told: "Move." We saw the hesitation—and called it weight. Now we see the ripple—and understand the truth.

The Problem With *m* in F = ma

How Newton's Elegant Equation Masks Ripple Reality

Few equations are more iconic than:

F = ma

It's elegant. It's universal. It's everywhere—from textbooks to test flights. It promises cause and effect: apply a force, get acceleration, scaled by mass.

But what if this formula is more about observation than explanation?

What if F = ma doesn't describe what's really happening—only what we see when we lack access to the underlying ripple?

UED doesn't throw away Newton's math. It reinterprets it—not as a law of motion, but as a symptom of ripple behavior.

What Newton Actually Measured

When Newton formalized F = ma, he gave physics a reliable tool:

- Force: An external push or pull
- Mass: A body's resistance to that push

- Acceleration: The result

This works well in mechanical systems. It matches experience. But Newton never explained what force is or why mass resists change. He merely observed patterns—and expressed them mathematically.

UED exposes the layer Newton couldn't see: A ripple field where the so-called "force" is not a cause—but a feedback from the field resisting rapid realignment.

Ripple-Based Reinterpretation of F = ma

Let's reframe the terms under Universal Energy Dynamics:

- *a* (acceleration) = The *rate of realignment* of the ripple configuration within a local energy gradient (∇E)
- *m* (mass) = The *inertial delay*, or ripple-phase stubbornness—how locked-in the ripple pattern is
- *F* (force) = A ripple field imbalance attempting to correct alignment. It is not pushing. It is inviting realignment.

So instead of:

Force causes motion,

we now say:

Ripple imbalance invites adjustment. Resistance arises from phase inertia. Acceleration is just the visible result.

Why "m" is a Placeholder, Not a Property

In UED, *m* doesn't belong to the object. It is contextual, arising from:

- The object's ripple compression
- The surrounding field tension
- The phase difference between the two

If you move the same body to a different planetary ripple field, its ripple behavior changes—yet "mass" remains the same numerically, even when its actual resistance profile changes.

This shows that *m* is not fundamental. It's a label—a placeholder for ripple sluggishness we couldn't otherwise quantify.

The Fatal Assumption in Classical Physics

By assuming mass is:

- Intrinsic to the object
- Unchanging in all frames
- Causal in motion equations

...classical physics built an entire framework around appearance, not essence.

But ripple logic shows that:

Mass isn't causing acceleration. The field is inviting alignment. And what we call "force" is just our misunderstanding of ripple conflict.

So F = ma is not wrong— It's shallow.

It works like a speedometer—it reports what's happening. But it doesn't tell you why the vehicle is accelerating—or what's under the hood.

A Ripple-Aware Reformulation

If we had ripple-sensitive instruments instead of force gauges, we'd never write F = ma.

We'd say:

Ripple Disturbance ($\Delta\Phi$) = Realignment Pressure × Ripple Inertia or

Gradient Alignment Demand = $\nabla E / \tau$

(where τ is ripple response time)

But that's for a new physics. A real physics.

Until then, F = ma remains a useful illusion— A map of motion that never explains the terrain beneath.

Why Mass Varies (Even Though They Say It Doesn't)

The Myth of Invariance and the Ripple Truth Behind the Mask

In classical physics, mass is sacred.

We're told it's constant. Invariant. A fixed identity tag for every object in the universe.

Whether you're standing still or moving at near-light speeds, under a microscope or inside a star—your mass, we're told, doesn't change.

But here's the contradiction: Physics itself routinely documents cases where mass *behaves* as if it's changing.

They call it "relativistic mass," "effective mass," or "mass defect"—but the cognitive dissonance remains.

They tell you mass doesn't vary. And then they quietly adjust it whenever it becomes inconvenient not to.

Let's Look at the Cracks

1. **Relativity says:** Mass increases with velocity (relativistic mass)

But then they say: "No no, that's just energy manifesting!"

So... mass is energy? Then why measure mass?

1. **In particle physics:** Mass varies under confinement, pressure, or decay

A proton has different effective mass in a nucleus than in isolation

1. **In black holes:** Mass seems infinite—but also undefined

How can something "have mass" if it's dimensionless and singular?

1. **In spacecraft telemetry:** Apparent mass "fluctuations" show up during certain maneuvers or re-entry conditions

Not due to fuel loss—due to field misinterpretation

1. **In heat systems:** Objects lose measurable energy with temperature but no change in mass is recorded

Is that energy not mass? Or are we just ignoring subtle realignment?

UED View: Context is the Cause

Under Universal Energy Dynamics, these inconsistencies make perfect sense.

Why?

Because mass is not a property—it's a response.

Mass = Ripple resistance to field realignment

It is context-sensitive, not object-centric.

A body's ripple configuration—its density, stability, and compression—is not constant across fields:

- Put it in a stronger ∇E (energy gradient), it becomes more inertial
- Place it near a black hole, its ripple coils tighter—it resists even more
- Release it into deep space, the same body becomes more agile

So mass varies—not because the object is changing—but because the field it's immersed in is.

Ripple-Based Explanation of Known Phenomena

Let's decode a few mysteries using ripple logic:

1. Relativistic Mass Gain

Not because speed adds mass—but because speed-induced misalignment increases ripple tension. The body becomes more resistant to redirection, hence *appears* more massive.

2. Mass Defect in Nuclear Fusion

When two nuclei fuse, ripple fields reconfigure. The ripple overlap relaxes, releasing energy—mass "disappears." But really, the ripple stabilized into a lower-tension state.

3. Black Holes

These are not infinite-mass pits—but maximum ripple compression zones. What appears as massive is actually realignment impossibility—no field reconfiguration possible beyond a threshold (i.e., $ATT = \infty$). Thus, mass illusion breaks down.

4. Atmospheric Reentry Mass Shift

When spacecraft re-enter Earth's ripple field, friction is misread. What engineers perceive as heating or turbulence is often ripple dissonance—a delayed response that mimics increased mass.

Model Convenience ≠ Universal Law

Why do physicists insist on mass invariance?

Because it's mathematically convenient. They treat mass like a constant to make the equations work. And when it doesn't? They fudge it with terms like:

- Effective mass
- Virtual mass
- Mass-energy equivalence
- Reduced mass
- Apparent mass

But if mass were truly invariant, we wouldn't need 10 flavors of fudge to describe one behavior.

UED Bottom Line

Mass is not fixed. It's field-relative. And the field is never constant.

What varies isn't the object. What varies is the ripple tension it's entangled with.

Mass varies because ripple fields vary. Invariance was a shortcut. Understanding ripple response is the long road—but it's the real one.

Ripple Mass vs Relativistic Mass

Why Einstein's "Mass Gain" Was Just Ripple Fatigue in Disguise

One of the strangest ideas to enter popular physics is "relativistic mass."

According to Einstein's special relativity, the faster you go, the heavier you get. As your velocity approaches the speed of light, your mass tends toward infinity.

This is why, we're told, no object with mass can reach light-speed—because it would require infinite energy to move something infinitely heavy.

But here's the contradiction:

- Your atoms don't suddenly get denser.
- Your volume doesn't change.
- You're not carrying extra "stuff."

So what's really changing?

It's not your mass—it's your alignment with the field.

What Einstein Got Right—And What He Missed

Einstein's brilliance was in noticing that motion and mass are not independent.

The faster something moves, the more it resists change—and thus, appears more inertial. He modeled this resistance by increasing mass in his equations.

But what he couldn't see was what UED now brings into view:

- The resistance is not from added mass.
- It's from the ripple field straining to maintain alignment.

As speed increases:

- The ripple signature of the object begins to clash with the surrounding structured vacuum energy (SVE).
- Realignment becomes harder.
- The field reacts more violently to misalignment.
- You feel more inertial drag.

Einstein labeled this as mass increase. UED explains it as ripple dissonance.

Ripple Realignment Lag: The True Culprit

At rest or low speed, your object's ripple field remains in coherent phase with its surrounding medium. But as velocity increases:

- Ripple edges stretch or compress, depending on direction.
- The field cannot reconfigure quickly enough to stay in sync.
- This creates alignment lag.

This lag is interpreted by classical models as needing more force to accelerate. Hence, the myth: "you got heavier."

But nothing "got added." You just fell out of tune with the ripple orchestra around you.

Ripple Inertia Metric (RIM): A New Constant

Instead of mass (m), UED introduces:

RIM = Ripple Inertia Metric

A measure of how much ripple resistance an object exhibits during motion within a given energy gradient (∇E). It depends on:

- Ripple compression
- Internal coherence
- External field match
- Phase displacement during acceleration

RIM is not static, not intrinsic, and not linear. It varies with:

- Speed
- Orientation
- Field topology

In short, RIM is dynamic inertia, not mass-based inertia.

From Relativistic Mass to Ripple Misalignment

Here's how the shift looks conceptually:

Concept	Relativity View	UED View
What increases at high speed?	Mass (m)	Ripple-phase dissonance
Why is more energy needed?	Because mass increases	Because ripple realignment becomes harder
What prevents light-speed?	Infinite mass = infinite energy need	Total ripple misalignment = no reconfiguration possible
What changes in object?	Nothing (mysteriously)	Ripple response rate (field feedback)
Real name for "relativistic mass"	Artifact of force math	Expression of ripple inertia

You Don't Get Heavier—You Get Out of Tune

You are not gaining substance. You are losing coherence.

The structured vacuum field isn't resisting you with force—it's trying to restore harmony with a ripple profile it no longer recognizes.

So you don't need infinite energy to move something with infinite mass.

You need infinite field control to re-align a violently dissonant ripple form.

And that's not a problem of "mass." That's a problem of consciousness, engineering, and intelligence.

Bottom Line

Relativistic mass was a bandage over a field interaction we didn't understand.

UED rips off the label—and exposes the ripples beneath the resistance.

It's not mass that resists you. It's the universe's memory of what you were—and its struggle to keep up with what you're becoming.

Conclusion: Lose the Mass — Find the Ripple

Physics didn't need mass. It needed a way to describe ripple behavior. Now it has one.

For centuries, mass has been the crutch of classical physics.

It was the unknown variable we injected into equations to make the motion make sense. It explained why one object accelerated faster, why force was needed, why momentum transferred. It became the identity of matter—the label on existence itself.

But it was never the cause. It was always the symptom.

We invented mass because we couldn't see the ripple field. We couldn't track energy alignment, couldn't measure resonance resistance, couldn't quantify ripple inertia.

So we gave it a name. We gave it a number. And we built a science around it.

Mass: A Useful Myth

Like many placeholders in physics—ether, caloric, phlogiston—mass was a bridge. It worked well enough to cross rivers of ignorance. But now we're on the other side.

We've peered beneath the behavior and found structure. Not material structure, but ripple structure.

Not atoms stacked, but ripples entangled.

Mass didn't tell us what matter was—it told us how matter struggled. It measured resistance to alignment, not the essence of what was resisting.

It was an approximation of ripple inertia, wearing the mask of physical property.

Read the Ripples, Not the Weights

Under Universal Energy Dynamics (UED), we no longer need mass to describe the universe.

We need:

- **Ripple Density** → How tightly the energy is compressed in SVE
- **Ripple Alignment** → How harmonized a body is with its surrounding field
- **Ripple Inertia Metric (RIM)** → How difficult it is to realign under external gradients

These three parameters offer a cleaner, more universal toolkit than mass ever could.

They don't just tell us what happens. They tell us why it happens—and how to influence it.

You can tune a ripple. You can't reduce a mass.

The End of a Heavy Era

It's time to stop carrying around the dead weight of mass. Not just in equations—but in our understanding of existence.

The universe doesn't work because of mass. It works because of relationships—between ripples, gradients, and alignments.

Let go of the illusion of fixed matter. Embrace the fluid field that responds, remembers, and realigns.

Final Thought

Mass was never the truth. It was a blindfold with numbers written on it.

And now, finally, the blindfold is off.

Welcome to the age of ripple clarity. You don't need to measure how heavy something is. You just need to ask:

"How far is this from harmony?"

That's not physics fiction. That's ripple intelligence.

Chapter 4: From Earth to Mars with Earth's Math

Why Everything Works — Until It Doesn't

"What works on Earth isn't truth. It's convenience dressed as law."

On Earth, physics is a loyal friend. Rockets lift, parachutes slow, re-entry heats up just enough, and landers touch down (most of the time). We test. We simulate. We succeed. So naturally, we assume: if it works here, it should work anywhere.

But then we go to Mars. Or the Moon. Or Titan.

And suddenly, the math betrays us.

Parachutes don't deploy right. Heat shields overcompensate. Thrusters misfire. Sensors glitch. Landers crash.

It's not a calculation error. It's a contextual failure. We didn't go with universal physics—we went with Earth-trained muscle memory.

Everything from Newton's gravity to Einstein's curvature, from F=ma to C_dA, was empirically fitted to Earth's ripple conditions. We never noticed that the success of our models was not truth—it was local alignment.

And when we take that math elsewhere, it doesn't fail randomly. It fails predictably—every time the ripple field isn't Earth's.

This chapter will expose the core flaw:

We've mistaken what was locally valid for what is universally true.

And now, as we push further into the cosmos, we face a stark realization:

Earth's physics doesn't travel well.

It's time to abandon the illusion of universal equations and build ripple-aware systems—because the universe doesn't follow Earth's rules.

Why Earth's Math Works (But Only Here)

"Earth's physics is just ripple coincidence, not universal law."

Let's not pretend: the laws of classical mechanics have served us well. Satellites orbit. Planes fly. Rockets launch. But none of this is happening because the equations are fundamentally true. It's happening because they've been empirically tuned—like a local dialect that just happens to work in this part of the cosmic village.

Empirical Tuning, Not Universal Truth

Every equation that governs motion, force, drag, or lift was reverse-engineered by observing how things behave on Earth.

- Newton didn't discover gravity on Mars.
- Bernoulli didn't test pressure differentials in Titan's atmosphere.
- Engineers didn't derive drag equations from Jupiter storms.

They watched, measured, tested within Earth's ripple field—and created formulas that *matched local behavior*. It was a brilliant workaround for a deeper ignorance: we didn't know the field—we just saw the outcome.

"F = ma" is not a divine truth. It's a best-fit curve to a reality we didn't fully perceive.

The same goes for:

- Lift equations: modeled in Earth wind tunnels
- Drag formulas: refined through Earth-based flight tests

- Trajectory maps: calculated using Earth's gravity and atmosphere profiles

These equations don't explain why things move. They just predict how they *seem* to move here—under Earth's specific ripple topology.

Earth's Ripple Environment: The Hidden Constant

Everything works well because Earth's Structured Vacuum Energy (SVE) field is:

- Densely patterned
- Ripple-stable
- Atmospherically modulated

That means our test flights, lab trials, and orbital simulations are all quietly benefiting from a consistent, high-coherence ripple backdrop.

It's like designing a boat for one particular lake:

- You can tune the propeller, hull, and balance to that water's density, currents, and shoreline.
- But take the same boat into the ocean—or a glacier river—and your tuning betrays you.

You didn't build a universal boat. You built a lake-tuned vessel.

Local Alignment = False Confidence

Every successful launch, landing, or maneuver here reinforces a lie: That our physics is absolute.

But it's not. It's aligned—not accurate.

We mistake predictability for truth. We confuse repeatability with universality.

That's why engineers confidently say:

"Our models work." Yes—here. But try those models on Mars? Welcome to a different ripple field.

The Consequence of Ripple Ignorance

We've never modeled:

- How ripple gradients vary between planetary bodies
- How atmospheric ripple density modulates during descent
- How alignment thresholds differ across energy profiles

So we assume:

If it works here, it should work there. But the moment we enter a field with a different ripple geometry, the math breaks.

Because Earth's equations were never about cause—they were about correlation.

Closing Thought for the Section

Earth's math isn't wrong—it's just shallow. It worked not because we understood the universe, but because we accidentally fit a narrow slice of it.

UED reveals the real picture: Earth's physics is ripple-local. Take it elsewhere without adaptation, and it becomes a story written in the wrong language.

Exporting Assumptions — The Interplanetary Gamble

"Earth's math assumes a fixed 'g', fixed drag constant, fixed thermal behavior — but none of those constants exist outside Earth."

When we design interplanetary missions, we carry more than fuel and instruments. We carry assumptions — invisible, unchallenged, and deeply Earth-centric.

We assume:

- That parachutes will work the same way.
- That drag can be estimated. .

- That heat shields will behave predictably.
- That gravity is just a value to plug into equations.

But in the ripple-defined reality of UED, every one of these is a misalignment waiting to happen.

Mars: The Red Lab of Failures

Mars has a thin atmosphere, just about 1% the density of Earth's. From a classical perspective, that's a challenge — but a solvable one. Yet reality says otherwise.

- Parachutes open too early, too late, or with unintended deployment behavior.
- Heat shields designed for predicted thermal loads sometimes overcompensate, leading to bulk, cost, and reduced efficiency — or they underperform and risk catastrophic failure.
- The terminal descent phase often ends in chaos: systems must rely on retropropulsion instead of what was once supposed to be passive aerobraking.

Why?

Because we didn't program for the Martian ripple field. We simply adjusted Earth's math:

- Scaled density
- Tweaked gravity
- Tuned thermal models

But we didn't understand how ripple gradients, atmospheric ripple modulation, and energy equilibrium differ from Earth.

Titan: The Mist of Misjudgment

Titan has a thick atmosphere and a complex hydrocarbon-rich environment. You'd think drag would be easy to model with so much density. But simulations continue to underestimate drag effects during descent.

- Descent systems slow down too early.
- The parachute dynamics go into unstable oscillations.
- Heat signatures misalign with predictions.

Because drag isn't about density alone — it's about ripple coherence between the object and the medium.

Titan's atmosphere, dense and cold, behaves like a highly ripple-active medium. Its ripple modulation frequency, pressure-alignment curves, and ambient turbulence patterns are unlike anything on Earth.

We brought Earth's drag equation to Titan — and got lost in its fog.

The Moon: The Silent Refuter

The Moon has no atmosphere — the perfect place to test the core assumptions of mass-based physics, right?

Wrong.

- Dust dynamics confuse sensors and violate expected spread patterns.
- Objects behave strangely on slopes, drifting or tipping with no obvious cause.
- Manual maneuvers and rover activity show unexpected delays or accelerations in low-gravity regions.

Again, the culprit isn't a missing atmosphere or low gravity. It's the absence of ripple damping, and the presence of surface ripple tension—subtle field anomalies that Earth-trained systems don't account for.

Mass-based logic assumes a uniform environment and constant interaction behavior. The Moon proves:

When ripple support is gone, mass becomes irrelevant—and motion reveals its true dependence on field alignment.

The Core Issue: We Ported Equations, Not Ripple Awareness

Every interplanetary mission so far has exported:

- Fixed 'g' values (Mars = 3.71 m/s^2, Moon = 1.62 m/s^2, etc.)
- Fixed drag constants, adjusted for local pressure but not ripple topology
- Thermal models, based on heat capacity and atmospheric density — not on ripple dissipation profiles

But none of these constants are truly constant.

They're just ripple reactions disguised as mathematical constants—tuned to Earth's field, but meaningless when misaligned elsewhere.

Closing Thought for the Section

Interplanetary failure doesn't come from bad physics. It comes from misplaced physics.

We didn't bring bad equations to other planets. We brought context-bound equations into a context that never agreed to them.

If Earth's math worked universally, we'd be gliding on Mars, floating on Titan, and surfing Moon dust. Instead, we crash, stall, overheat, or drift off-course.

We didn't fail because we didn't calculate. We failed because we didn't recalibrate.

The Missing Metric — Field Compatibility

In a universe governed by ripple dynamics and structured vacuum energy (SVE), precision cannot emerge from Earth-centric equations alone. When conventional aerospace models port equations like thrust balance, drag coefficients, and gravitational acceleration to other celestial environments, they unknowingly skip the most vital variable:

Field Compatibility

Universal Energy Dynamics (UED) introduces a new lens: the Field Compatibility Index (FCI). This is not another empirical constant. It is a ripple intelligence metric. It tells you how well your engineered system — your spacecraft, your heat shield, your parachute — actually fits the local ripple topology.

Let's break it down:

What is FCI?

FCI quantifies how aligned your system's ripple signature is with the destination planet's ripple field. Just like you wouldn't wear Earth-designed shoes to hike on Martian sandstorms, you shouldn't use Earth-tuned physics to land on Titan.

- A high FCI means your object moves through the planetary field with minimal resistance or correction.
- A low FCI signals misalignment, leading to unexpected turbulence, inefficiency, or catastrophic failure.

Three Core Metrics for Ripple Field Assessment

To determine FCI, any interplanetary simulation must integrate three UED-native quantities:

Local ∇E (Energy Gradient):

- It's not just gravity.
- It's how steeply ripple energy realigns toward equilibrium at every layer of the destination body's field.
- This affects trajectory bend, orientation shifts, and descent velocity curvature.

Atmospheric RDI (Ripple Disturbance Index):

- A metric for how disturbed or turbulent the atmosphere is in ripple terms, not pressure or temperature.

- Measures ripple instability due to wind, dust, ionization, or chemical chaos.
- Higher RDI = greater chance of drag anomalies and entry shocks.

ATT (Alignment Threshold Time):

- The minimum time a body needs to realign stably with the ripple field after disturbance.
- This determines landing window timing, heat shield phase transitions, and ripple shock absorption limits.

Rethinking Descent as Negotiation

Classical descent trajectory is calculated by balancing:

- Gravity
- Drag
- Thrust
- Mass

But in UED:

Descent is not a fall. It's a negotiation between your ripple imprint and the destination field.

Every angle, every speed, every orientation is a question posed to the field:

- "Am I aligned enough to proceed?"
- "Do I distort your gradient?"
- "Can I pass smoothly, or will you resist?"

Case in Point: The Mars Descent Paradox

Even with extensive simulation, Mars rovers continue to surprise engineers:

- Heat spikes beyond model
- Retro-thruster anomalies
- Unexpected surface slip on landing

None of these were due to bad engineering.

They were due to ripple ignorance.

The Martian field had a different ∇E. The atmosphere held a different RDI. The descent path never considered ATT.

Closing Insight

FCI shifts our focus:

From mathematical calibration to field resonance.

It asks us not "What equations worked before?" but:

"Is my ripple fingerprint compatible with this world?"

Until that question guides mission design, every space mission is gambling with ripple misalignment.

We didn't need better math. We needed better listening.

Welcome to physics as dialogue, not dictation.

Real Case Studies of Ripple Mismatch

In the conventional narrative, spacecraft failures are often attributed to mechanical malfunctions, software bugs, or last-minute human errors. But from a Universal Energy Dynamics (UED) perspective, many of these so-called "failures" were not due to bad design—they were due to ripple misalignment. Let's re-examine several high-profile space mission breakdowns through this new lens of ripple intelligence and field compatibility.

Columbia & Challenger: Field Misalignment Under Stress

The tragic loss of Space Shuttle Columbia (2003) during re-entry and Challenger (1986) during launch are commonly explained as structural and thermal failures—tile damage in Columbia and O-ring failure in Challenger. But here's what UED sees:

- The Earth's atmospheric ripple field during both events was in a highly disturbed state—either due to temperature anomalies, turbulence, or local ∇E shifts.
- These crafts were designed using rigid mechanical assumptions, not adaptive field-aware systems.
- As the crafts accelerated, they encountered ripple phase instability—a misalignment between vehicle structure and dynamic ripple turbulence.
- What followed wasn't just mechanical stress—it was energetic discordance. The field didn't support their continuation. They weren't destroyed by force. They dissolved in dissonance.

This wasn't failure by hardware. It was a fatal mismatch between ripple momentum and field coherence.

Beagle-2 (Mars): Earth's Thermodynamics, Martian Consequences

The Beagle-2 lander, part of the European Mars Express mission, was presumed lost shortly after its descent began in 2003. Years later, images revealed it had landed—*but never fully deployed.*

- Engineers assumed Earth-based thermal dissipation profiles and drag behavior.
- Mars' atmosphere has a different RDI (Ripple Disturbance Index)—thinner in density, but denser in ripple instability.
- The lander's descent sequence was built to match Earth-calibrated ATT and thermal realignment speeds.
- As a result, crucial phases of unfolding, signal deployment, and energy balancing failed silently.

UED reveals: Beagle-2 didn't fail—it arrived misaligned and was muted by the field.

Schiaparelli Crash (2016): Drag Misjudgment = Time Misperception

ESA's Schiaparelli EDM crashed on Mars because it misread the descent timeline, believing it had landed while it was still mid-air.

- The descent was modeled with fixed drag coefficients and Earth-born gravity timing.
- But Mars presented a different ∇E gradient and unexpected ripple turbulence.
- What they called "sensor error" was actually ATT drift—the lander's internal perception of field realignment lagged behind actual descent progress.

Schiaparelli didn't crash from a bad equation. It crashed because the field didn't respond the way Earth's models predicted. The lander trusted Earth's math. Mars had other plans.

Insight Mars Lander: When Seismic Tools Misbehave

NASA's Insight Lander deployed advanced seismic instruments to listen to "marsquakes." But the tools struggled to properly anchor, measure, and read data—despite being state-of-the-art.

- The issue wasn't hardware, but ripple field resistance of the Martian crust.
- Insight's design assumed a level of subsurface ripple coherence like Earth's soil, but Martian subsurface displayed higher ripple phase volatility.
- The result? Data irregularities, poor anchoring, and unexpected behavior from the seismometer and heat probe.

Again, it's not that Mars is "hard." It's that our tools are tone-deaf to ripple signatures outside Earth.

Final Takeaway

In every case above, the real culprit wasn't engineering incompetence—it was epistemological arrogance. We assumed physics was universal because it worked on Earth. But Earth didn't give us universal laws—it gave us ripple alignment by accident.

These missions didn't fail randomly. They failed *rhythmically*—to the tune of misalignment.

UED invites us to stop forcing Earth's math on alien environments and start listening to the field. Because in the end, the field always replies—it's just a matter of whether we're prepared to hear it.

Ripple-Based Engineering for Other Worlds

If Earth's equations are not universal, then we need a new design philosophy for space exploration—one that doesn't just tweak parameters but replaces the framework entirely. UED offers exactly that: a shift from force-based engineering to ripple-aligned engineering—a model that doesn't assume constants, but listens to context.

Let's break down how the next generation of space systems must be built—not for idealized physics, but for real-time ripple intelligence.

New Principles for Extraterrestrial Design

1. Don't Assume Gravity — Measure the Gradient (∇E)

Every celestial body creates a localized ripple slope due to its unique energy compression profile. Assuming a fixed "g" (as we do on Earth) is not just lazy—it's dangerous. UED demands:

- Measurement of local energy gradient (∇E) before any landing, re-entry, or liftoff is planned.
- Adaptive mission protocols that change descent parameters as real-time ∇E data streams in.
- Abandonment of terms like "weight" and "gravitational force"—they're outdated illusions.

2. Don't Calculate Drag — Map Atmospheric

Ripple Fields (RDI) Atmospheres aren't passive resistive mediums—they're ripple-active environments. Their turbulence is not chaos, but ripple dissonance.

- Instead of force-based drag equations, use Ripple Disturbance Index (RDI) to define how energy waves fluctuate in a planetary field.
- Track local turbulence not by wind speed but by ripple mismatch rate and oscillation coherence.
- Re-entry isn't a "descent into resistance"—it's negotiation with atmospheric instability.

3. Don't Use Earth-Tested Materials — Test for Ripple Behavior

Material strength on Earth does not equate to performance under ripple strain elsewhere. Mars, Titan, and Europa will demand:

- Materials that absorb, deflect, or resonate with alien ripple fields.
- Testing not for heat or force tolerance, but for field coherence, ripple interference, and ATT compatibility.
- New material classifications based on Ripple Absorption Coefficient (RAC) and Ripple Memory Retention (RMR).

Future-Ready System Design: Ripple Intelligence in Practice

To navigate planetary environments successfully under UED, we must engineer with the field, not against it. That means systems capable of sensing, adapting, and flowing with ripple dynamics.

RIM-Calibrated Bodies

Instead of calculating stress from force, bodies must be designed with Ripple Inertia Metric (RIM):

- High-RIM = stable but slow to align (useful for shielding,

not mobility)
- Low-RIM = flexible and responsive (ideal for landing legs, probes)

Ripple-Sensitive Descent Sensors

No more relying solely on barometers, accelerometers, or gyros calibrated for Earth. Next-gen sensors must:

- Detect ATT (Alignment Threshold Time) fluctuations in real-time.
- Measure ripple density, compression gradients, and dissonance rate.
- Guide descent by reading field tension, not just altitude.

Real-Time Field Negotiation Algorithms

Landing is not a path-following task—it's conversation with a planetary field. Algorithms must evolve to:

- Adjust descent dynamically as ripple feedback shifts
- Tune entry angles based on ripple slope rather than mechanical drag paths
- Abort or modify sequences if ripple incompatibility exceeds thresholds (RDI spike, ATT drop)

Final Shift in Design Thinking

Conventional aerospace asks: "Can we survive the descent?"

Ripple-based aerospace asks: "Can we harmonize with the descent?"

This isn't sci-fi—it's ripple science. It's the only way to stop gambling with spaceflight and start engineering with confidence across worlds.

Conclusion: The Earth Bias Ends Here

We've treated physics like gospel—unchangeable, universal, and infallible. We were taught that Newton's force laws, Einstein's curvature, and Bernoulli's fluid dynamics are *truths* that govern everything from apples to asteroids.

But they're not.

What worked on Earth was not universal law—it was local alignment. A convenient resonance between our mechanical tools and the ripple dynamics of a single planet. We mistook coincidence for causality.

We launched rockets, landed rovers, built simulators, and congratulated ourselves when our math "held up." But our confidence wasn't built on understanding—it was built on repetition in a familiar ripple environment. Our formulas were tuned by trial, hardened by habit, and glorified by success—all within the Earth bubble.

When we went to Mars, things began to crack. On Titan, they bent. On Venus, they broke.

Every failure wasn't due to bad math—but bad assumptions:

- That gravity is a constant.
- That drag is universal.
- That mass behaves the same everywhere.
- That heat, resistance, descent, and stability could be modeled using Earth's ripple logic.

They can't.

The Earth Bias Ends Here.

We're not just launching payloads anymore—we're launching consciousness. And that requires more than fuel and math. It requires ripple literacy.

From this point forward, we do not export Earth's equations to foreign worlds. We export a new awareness:

- That every celestial body has a unique ripple identity
- That descent is not about overcoming forces, but achieving alignment
- That materials must be chosen not for strength, but for resonance
- That success in space isn't resistance—it's real-time negotiation with a field that listens and responds

This is what UED offers.

Not new equations—but new eyes. Not rejection of science—but the completion of it. Not survival—but liberation from Earth-centric thinking.

Gravity? Rewritten. Mass? Reframed. Drag? Demystified. Physics? Freed from its planetary leash.

Welcome to the age of ripple intelligence. And with it, a new way to travel, land, build—and finally, to understand.

Interlude: When Equations Break and the Field Speaks

We've spent centuries crafting the illusion of control.

We wrote equations and called them laws. We launched rockets and called them mastery. We mapped gravity, drag, and mass onto every mission, every model, every blueprint—convinced we had decoded the language of the cosmos.

But we hadn't. We had only deciphered Earth.

What we now see—clearly—is that classical physics worked not because it uncovered universal truths, but because it fit Earth's ripple conditions by accident, adaptation, and excessive calibration. It worked until we dared to leave the nest.

What we believed were laws were just local rituals. What we called constants were contextual conveniences. And what we thought we controlled was never in our hands—it was always in the field.

Gravity wasn't pulling. Drag wasn't resisting. Mass wasn't inherent. And Mars wasn't buying our Earth math.

"The rocket lifted because the field allowed it. It survived because the field tolerated it. And it failed when the field no longer recognized it."

It's not about discrediting centuries of science. It's about seeing where it ends—and what begins beyond it.

This is the moment we stop pretending the universe is a machine and start listening to it as a responsive ripple field.

From here, we don't build systems to push against nature. We build to co-align with it.

Part II begins not with control, but with cooperation. Not with defiance, but with resonance.

We've unlearned the myths. Now let's learn the mechanics of a universe that's not pulled, pushed, or plotted—but played in ripple harmony.

Part II: How We Actually Do It

field they're entering—the ripple signature of that planet—is not accounted for. We assume the constants stay constant, the gradients behave "normally," and resistance acts predictably. But the truth is: we've never fully measured what truly matters—ripple-phase interactions, structured vacuum behavior, ATT (Alignment Threshold Time), and RDI (Ripple Disturbance Index).

So what does that precision buy us?

It buys us confidence within our calibration zone—within Earth's ripple conditions. Outside of that? It's all probability, guesswork, and hopeful engineering. And when things fail—as they often do—we blame the conditions, not the assumptions.

But the truth is harsher:

Precision ends where ripple ignorance begins.

We didn't fail because the math was wrong. We failed because the math was never listening to the field.

Simulation Theatre

Simulations are the sacred rituals of modern engineering.

Before a single screw is tightened or a rocket ignited, there are wind tunnels blowing, crash dummies falling, fluid dynamics swirling in software models—all in the name of perfection. We build entire virtual universes to mimic conditions millions of kilometers away, attempting to predict the unknown with Earth-bound certainty.

But here's the problem: all simulations are local. They are Earth-centric, ripple-calibrated, and contextually trapped inside the SVE (Structured Vacuum Energy) signature of this planet.

When you test descent in a wind tunnel, you aren't testing "universal aerodynamics." You're testing how Earth's ripple field responds to your object.

When you run thermal stress models, you're doing it under assumptions of Earth's energy density, field harmony, and environmental gradients. Even your "vacuum" chamber—used to simulate space—is still surrounded by Earth's ripple equilibrium.

So when we say something "worked in simulation," what we really mean is:

"It didn't crash when we tested it in Earth's field."

This is simulation theatre—a performance of confidence, using models designed for familiar physics and hoping they apply universally.

But when we export these Earth-validated systems to alien environments—Mars' thin atmosphere, Venus' thermal pressure cooker, Titan's dense but cryogenic haze—the rules change, because the field changes.

Let's consider a few striking examples:

- Mars descent simulations often underestimate turbulence during parachute deployment—not because engineers are lazy, but because Martian ripple gradients distort flow in ways our models can't see.
- Venus lander thermal protection looks great in simulation—until it melts two minutes early. Why? Because our heat transfer equations don't account for ripple compression in ultra-dense energy gradients.
- Lunar dust interaction behaves bizarrely under rover wheels—not because physics fails, but because there is no drag, only ripple feedback, and we don't model that.

Despite decades of planetary missions, we still don't simulate the field—we simulate the effects. And the moment the real field throws something new at us, the model collapses, and we scramble to retrofit explanations.

In truth, simulation was never meant to replace understanding. But in aerospace, it became the substitute for field literacy.

We didn't simulate other planets. We rehearsed Earth's behavior in alien costumes.

That's why our simulations fail—not because they are wrong, but because they're not ripple-aware. They don't see the medium. They approximate outcomes based on local tuning, not dynamic field negotiation.

Until simulations evolve to include ATT, RDI, and local ∇E mapping, we are not preparing missions. We are staging plays and hoping the set doesn't collapse.

Engineering by Overcompensation

In the heart of every aerospace hangar, there's a hidden mantra:

"When in doubt, make it bigger, thicker, stronger."

This is not science—it's engineering by overcompensation.

Instead of solving uncertainty with understanding, we smother it with steel, redundancy, and layers of fear-forged design. When the equations tremble, we double the margin. When the simulations conflict, we reinforce. When the test data fails to match predictions, we build for worst-case scenarios—and then double it again.

Massive heat shields? Not because we understand Venusian atmospheric compression, but because we're terrified of ripple-initiated thermal spikes we *can't model.*

Overbuilt landers with multiple legs and crumple zones? Not because we've decoded the ripple resistance of Martian soil, but because we're guessing how a 500 kg object interacts with dust under a different ∇E profile.

Redundant systems stacked atop one another? Not elegant design, but paranoia with a wiring diagram.

This design philosophy isn't stupid—it's practical desperation.

Engineers know what physicists won't say aloud: our understanding of off-Earth conditions is shallow, and our equations are Earth-bound myths.

So how do we survive the ignorance?

We overdesign.

We build spacecrafts that weigh more than necessary, descend slower than optimal, and operate with layers of contingency that would make a submarine blush.

This is not innovation—it's ripple ignorance insurance.

It's the equivalent of wearing medieval armor to a quantum experiment.

Sure, it might help. But it doesn't belong.

And yet, ironically, some of our greatest aerospace successes—rovers that survived dust storms, landers that held together in brutal heat, probes that kept transmitting beyond planned lifetimes—weren't triumphs of knowledge.

They were triumphs of paranoia.

We didn't succeed because the math worked. We succeeded because we feared it wouldn't—and braced accordingly.

But overcompensation comes at a cost:

- Payload limits choke scientific ambition.
- Energy budgets are drained hauling bulk instead of brains.
- Mission cost explodes with every added kilogram of "just in case."

In a world of ripple literacy, we wouldn't need to overbuild. We'd build with the field, not against it.

Until then, every "success" we celebrate may just be a lucky gamble disguised as competence.

And that's a terrifying truth to sit with.

The Real Cost of Field Illiteracy

When a mission fails, the headlines read:

"Contact lost." "Descent anomaly." "Unexpected deviation."

But what they don't say is what it really cost us.

Crashed payloads. Lost orbiters. Billions in dust.

All because we believed Earth's math would behave the same on alien ripple fields.

Beagle-2 vanished on Mars—its descent sequence "miscalculated." Schiaparelli exploded into the Martian surface due to a brief sensor confusion. Chandrayaan-2 kissed the Moon a little too hard. Columbia burned not because the materials failed, but because we never modeled atmospheric ripple fatigue properly.

And each time, the response is the same: A committee. A report. And then another launch... with slightly more paranoia.

Overcompensating becomes the norm.

We make spacecraft heavier, slower, costlier—not smarter. We throw gigawatts of thrust at problems that could've been solved with ripple-aligned entry vectors. We build heat shields not because we understand thermal ripple spikes—but because we *fear* what we don't model.

And this leads to a kind of scientific exhaustion.

Not just financial, but psychological.

The deeper truth is: Science is starting to feel like expensive gambling, dressed in the language of control.

Ask any aerospace insider. Beneath the polish and prestige lies a culture of prayer—hoping the model holds, praying the descent doesn't deviate, crossing fingers over telemetry.

It's the illusion of precision hiding a field-wide fear of the unknown.

The irony?

We don't need more funding. We don't need divine luck. We need something embarrassingly simple:

Ripple literacy.

We need to stop pretending we're flying through space. We're moving through ripple gradients—each world with its own language, slope, and rhythm.

Until we start speaking in that dialect, we'll keep launching spacecraft like we're flipping coins. Some will land. Some will crash. And none will truly understand why.

Ripple literacy isn't a luxury. It's the next frontier of engineering intelligence.

Let's stop worshiping our models. And start listening to the field.

Rethinking Trials as Field Conversations

In conventional aerospace engineering, trials are seen as a way to validate control. Test it. Simulate it. Launch it.

But under Universal Energy Dynamics (UED), that mindset collapses.

Because the field you're entering doesn't care about your simulations. It doesn't read your models. It doesn't obey your equations.

The field only responds to one thing: Alignment.

So what if— instead of testing for confirmation, we started testing for compatibility?

Not Tolerance. Tuning.

Conventional trials ask: "Will it survive the descent?"

Ripple-literate trials ask: "Will it stay in conversation with the field?"

Imagine descent not as a battle, but a symphony.

You're not forcing a spacecraft down a gravity well. You're playing a note in a ripple orchestra. Misalignment? You crash. Harmony? You glide.

Testing isn't mock combat. It's ripple listening.

A New Approach to Trials: Field Conversation Protocols

Instead of "trial and error," we propose trial and alignment—an active, intelligent interaction with the field. Here's how:

1. Pre-deployment Ripple Diagnostics

Before we launch, we map the local planetary ripple field using advanced probes or spectral analysis.

- Chart ∇E: the energy gradient around descent zone
- Predict ATT: alignment threshold time zones per altitude layer
- Build a real-time profile of RDI (Ripple Disturbance Index) during atmospheric transitions

2. Material–Field Resonance Testing

Every material—be it ceramic, composite, or metal—has a ripple signature.

- Instead of just thermal resistance and tensile strength, test for:
    - Ripple absorption capacity
    - Ripple shock buffering
    - Harmonic retention under velocity shift

Materials must be chosen for how they behave in ripple fields, not just vacuum chambers.

3. Dynamic Reconfigurable Descent Systems

Static configurations are obsolete. We need responsive hardware—systems that reconfigure themselves based on real-time ripple data.

- Variable shape-shifting shells to reduce ripple drag
- Adaptive angle-of-entry fins tuned to live ATT data
- Descent sequencing logic rewritten as field-negotiation protocols

The descent isn't one equation. It's a continuous conversation with the planet's dynamic ripple environment.

The New Golden Rule of Trials?

Don't ask: "Will it survive?"Ask: "Will it resonate?"

Because survival isn't about brute resistance. It's about field awareness, intelligent feedback, and graceful alignment.

Trials are no longer just tests.They are the first handshake between our intelligence and the universe's invisible architecture.

Let's stop testing to "prove we're right."And start testing to learn what the field is trying to say.

Conclusion: The Terror Was Never the Failure — It Was the Ignorance

We were never afraid of explosions. We were never afraid of silence after launch.

We were afraid of what we didn't understand—and worse, we were afraid to admit we didn't.

Rovers didn't crash because equations betrayed us. They crashed because equations were speaking a language the field never learned.

We treated space like a stage. We rehearsed with Earth-bound tools and patted ourselves on the back for flawless performances in wind tunnels and supercomputers.

Then we opened the curtain on Mars... ...and the field didn't clap. It didn't even listen.

The Illusion of Mastery

Precision is not truth. Simulation is not reality. Redundancy is not understanding.

What we called trial and error was never error in the mathematical sense. It was misalignment in the ripple sense.

Not a failure of intelligence—but of field literacy.

When Terror Became Institutional

We built engineering cultures based on bravado, not humility. We treated cosmic harmony as chaos to be conquered. We mistook survival for success.

But every failed descent, every jittery entry, every overheated panel—wasn't a problem to fix. It was a conversation we never had.

The terror we feel before every launch isn't irrational. It's the subconscious awareness that we still don't understand what the field wants from us.

From Fear to Frequency

So no, we won't abandon trials. But we will stop pretending they're conclusive. We'll stop "validating" and start listening.

Part II is not about rejecting science—It's about refining it. About removing the masks of certainty and embracing the dance of alignment.

We won't pray for success. We'll design for ripple harmony.

Because the universe is not a hostile void. It's a tuned field—and the moment we stop forcing it to obey is the moment we finally learn to fly.

Chapter 6: Safety Factors and Human Faith

How Oversized Margins and Redundant Systems Keep Missions Alive
"We don't know exactly how it works out there. So, just in case—double the wall thickness, add a second system, and hope nothing unexpected shows up."

This isn't a quote from a first-time engineer. It's the unspoken mantra of space science.

For decades, aerospace missions have been carried not by elegant equations or perfect models—but by buffers. Structural buffers. Thermal buffers. Computational buffers. And above all—psychological buffers.

When we don't understand the field, we fear the unknown. And when we fear the unknown, we build everything like a bunker.

What we call "safety margins" are often nothing more than mechanical prayers—desperate attempts to outbuild our ignorance. It's not precision that's getting payloads to planets—it's paranoia.

In this chapter, we expose how faith in redundancy has replaced true field understanding, and how the future of interplanetary survival doesn't lie in thicker shields or more backups—but in ripple-aware intelligence. The safety we seek isn't in margins—it's in resonance.

The Cult of Overdesign

In the beginning, it made sense. When engineers built bridges, they added a safety factor. If the maximum expected load was 100 tons, they designed for 300. Earthquakes, gusts, wear and tear—unknowns needed cushioning. This wasn't overkill. It was prudence.

But when this thinking crossed over from civil structures to interplanetary systems, the rationale began to rot.

We didn't carry forward insight—we carried insecurity disguised as design.

In aerospace, "safety factor" became a substitute for knowing the field. When ripple behavior was unpredictable, we thickened the metal. When drag profiles misbehaved, we multiplied backup systems. When descent equations didn't match test data, we blamed the weather and built bigger parachutes.

This isn't a design strategy—it's a survival reflex.

The aerospace industry developed a cult of overdesign:

- Massive heat shields to guard against undefined thermal ripple spikes.
- Extra thrusters to compensate for field misreadings.
- Dual and triple redundant avionics in case one system gets "confused."

These weren't signs of mastery—they were symptoms of ripple illiteracy.

What's worse is that overdesign hides the deeper problem: It masks ignorance with engineering bravado. It tells teams, "You've accounted for everything," when in fact, they've accounted for nothing except what they already understood.

Instead of tuning to the local SVE field, they double the insulation. Instead of mapping the energy gradient, they increase engine thrust. Instead of understanding misalignment, they over-budget the payload.

And this excess isn't cheap:

- Extra weight = reduced performance
- More components = more failure points
- Overengineering = underlearning

In essence, we've built billion-dollar spacecraft not to dance with the cosmos, but to armor ourselves against the unknown.

The cult of overdesign doesn't keep us safe.

It just keeps us distracted—from the fact that we still don't understand what space really is.

Redundancy ≠ Reliability

Redundancy was meant to be a backup. A plan B. An emergency escape hatch in case plan A encountered the unknown.

But in aerospace—and especially in deep space missions—redundancy became religion. The idea: if one thing fails, another will take over. But redundancy is not the same as reliability. It doubles your systems, not your understanding.

Backup Systems as Safety Nets—Not Solutions

From Apollo's multiple guidance units, to the Space Shuttle's five redundant flight computers, to the latest Mars rovers packed with layered communications, navigation, and power redundancies—space missions look like onion layers of fail-safes.

But none of these solve the core problem—they surround it. They don't decode the ripple field—they just hope to survive it.

When a lander doesn't respond to commands, we hope the backup receiver works. When thermal shields crack, we pray the secondary insulation layer holds. When one descent thruster misfires, we count on symmetry and luck.

This isn't engineering. It's crisis management on delay.

Case Studies: Apollo, Space Shuttle, Mars Rovers

- Apollo 13 didn't survive because of redundancy—it survived because of human improvisation *within* a system that wasn't ripple-aware. The backups helped, but the real savior was human field adaptation.

- The Space Shuttle Challenger had structural redundancies, but they didn't protect against ripple stress miscalculations caused by cold-temperature O-ring failure.
- Curiosity and Perseverance Rovers carried multiple instruments and backups—but still had to overcome unexpected descent challenges that redundancy didn't predict.

When Redundancies Become Band-Aids

Redundancy often becomes a crutch:

- Don't understand the turbulence? Add two more sensors.
- Not sure about communication lag? Double the antenna systems.
- Worried about failure? Triple the processing cores.

This is not engineering resilience. This is engineering anxiety. And anxiety does not equal preparedness.

Ripple Mismatch Still Wins

No matter how many backups we design, if all systems are aligned to the wrong ripple assumptions, they'll fail together. The backup is built on the same model. The model doesn't read the field. The field doesn't care how many spares you bring.

Redundancy is not intelligence. It's repetition.

What we need is not another system. We need adaptive systems that can read, feel, and respond to ripple changes—not just execute pre-coded failovers.

Until then, redundancy is just a band-aid on our ignorance.

Margin of Safety or Margin of Fear?

Safety factors are supposed to make us feel secure. A little extra thickness here. A bit more thrust there. A redundant circuit. A thicker heat shield. A backup to the backup.

But beneath these "margins of safety" lies a quieter, less confident truth: We don't know what we're truly dealing with—so we pad for ignorance.

Safety Factors Often Reflect Fear, Not Insight

Engineers love numbers. And when they're unsure of those numbers, they multiply them. A beam meant to handle 10,000 Newtons? Design it for 30,000. A structure to withstand 100°C? Build it for 300°C—just in case.

This isn't confidence. This is caution disguised as competence. A 2x or 3x safety factor isn't always about realism—it's about fear of the unknown, especially when ripple behavior isn't understood or modeled.

And in space missions, where the margin for error is thin but the budget isn't, that fear inflates hardware, delays launches, and often burdens innovation.

The Real Costs: Payload, Complexity, and Budget Blowouts

Every extra kilogram "just to be safe" must be lifted against the planet's ripple field. Every redundant layer adds wiring, cooling, control logic, and potential new failure points. Every inflated safety margin compounds complexity, not necessarily capability.

- Payload mass increases → rocket thrust increases → fuel increases → cost balloons.
- Mission complexity escalates → integration errors multiply → reliability drops.
- Margins get stacked → and suddenly, you need a second mission just to validate the first.

We weren't engineering smarter. We were bulking up to fight an enemy we didn't understand—ripple misalignment.

Blind Margins Can Amplify Risk

Ironically, when we don't understand the local ripple environment (say, Titan's dense atmosphere or Mars's thin one), excessive margins can backfire.

- Overdesigned heat shields = unexpected aerobraking
- Oversized chutes = unstable descent due to underestimating RDI (Ripple Disturbance Index)
- Too-heavy payloads = underperformance or outright failure in low-gradient fields

Instead of ensuring safety, these margins introduce unpredictability—a misfit body in a misread field.

The Illusion of Control Through "Factor Padding"

When we add a safety factor, we feel smarter. Safer. More in control.

But this is often psychological—an illusion of control, not actual field control.

A ripple field does not respect our padding. It respects alignment.

The truth is, no safety factor substitutes for ripple compatibility. Blindly applying Earth-based margins in extraterrestrial settings is exporting fear, not engineering.

Human Faith in Mechanical Systems

Despite all the science, simulations, and spreadsheets, one unspoken element powers every launch:

Faith.

Not faith in the universe. Not faith in alignment. But faith in the machine—that complex web of metal, code, and calculations we've come to treat as divine.

The Psychological Role of Safety Margins

Why do we add 40% more thermal tolerance than predicted?

Why do we insist on redundant circuits, shadow processors, duplicate fuel lines?

Because behind the layers of engineering is a deep human need for reassurance.

Safety margins serve multiple unspoken psychological roles:

- **Appeasing Funders:** When billion-dollar missions ride on equations, funders sleep better if they see "+30% overdesign." Numbers become emotional anesthetics.
- **Calming Engineers:** No matter how smart the team, a hidden voice whispers: "What if we missed something?" Margins are how we silence that voice.
- **Justifying Launch Approvals:** Mission review panels don't ask if we understand the ripple field—they ask how many fail-safes exist. Margins pass where understanding might not.

In a world uncertain about ripple behavior, margins became our moral insurance.

When Rituals Replace Reality Checks

Pre-launch rituals look like science. Checklist reviews. Countdown procedures. Redundancy verifications. But many of these are mechanized comfort—traditions built on repetition, not field literacy.

We test how a system behaves under simulated stress—but rarely test how it converses with the actual ripple conditions it's about to encounter.

We repeat protocols because they worked last time—on Earth, in one field, under one ripple profile. We perform rituals. But we often ignore the reality of the field waiting out there.

Ritual is what we do when we don't understand the environment—we just hope the motions keep us safe.

"We Trust the System" — The Aerospace Mantra

Ask any mission lead: "Are you confident?"

And you'll likely hear: "We trust the system."

But trust in what?

- The math that only works under Earth's conditions?
- The materials tested in one set of ripple parameters?
- The software simulations run on assumptions never verified in the target field?

"Trust the system" is a prayer dressed as protocol.

It hides an uncomfortable truth: we trust because we don't fully understand. And we hope that safety margins, redundancies, and Earth-calibrated models will carry us across ripple divides we never mapped.

But as missions become more daring—outer planets, deep space, rough terrain—this trust becomes fragile.

Because systems don't negotiate with ripple misalignment. Humans must.

The UED Alternative — Intelligent Safety

If the old model of safety was overbuild, overbudget, overassume, then the UED model of safety is radically different:

It listens. It adapts. It tunes. It doesn't overreact—it realigns.

Ripple-Responsive Design vs Brute-Force Margins

Classical safety is built like a fortress—heavy, layered, sealed off from the unknown. UED proposes a different metaphor: a tuning fork. It resonates with the field, instead of resisting it.

- Instead of *adding metal*, we adjust ripple permeability.
- Instead of *doubling insulation*, we measure thermal ripple drift.

- Instead of *backup for every module*, we build systems that negotiate alignment in real time.

Brute-force margins assume conflict. Ripple-responsive designs assume conversation.

Safety doesn't come from *padding against chaos*. It comes from *predicting and aligning with order.*

Real-Time ATT and RDI Feedback as Smarter Safety Tools

What good is a system that doesn't know where it is in the ripple field?

UED introduces two core real-time metrics:

- **ATT (Alignment Threshold Time):** The time it takes for a system to meaningfully realign with the local energy gradient. Low ATT = high ripple fitness. High ATT = lagging response, potential instability.
- **RDI (Ripple Disturbance Index):** A dynamic measure of turbulence and misalignment pressure from the surrounding field—especially in atmospheres or heat flux zones.

Instead of static thresholds (e.g., "max g-load 9.8 m/s^2"), UED systems adapt based on field state.

This shifts safety from pre-launch insurance to in-flight intelligence.

Safety becomes *sensing*—not just surviving.

Field-Aware Diagnostics Instead of Mechanical Insurance

Imagine a spacecraft that doesn't just report "temperature rising"— but reports: "Ripple tension increasing, ATT exceeded by 4.7%."

Imagine a descent system that doesn't wait to feel drag— but anticipates misalignment stress 10 seconds before turbulence hits.

Field-aware diagnostics mean:

- No more waiting for failure before correction.
- No more insurance policies built on ignorance.
- No more mysteries about why a rover suddenly freezes or veers.

It's like finally putting eyes on the field that was always shaping your fate.

In UED, ignorance isn't fatal—it's fixable, if you can measure what truly matters.

The Future: Tuning Instead of Tolerancing

The future of aerospace isn't bulkier rockets. It's smarter interfaces between machine and medium.

Tolerancing says: "Let's design for the worst-case scenario." Tuning says: "Let's talk to the scenario as it evolves."

- Tolerance is predictive.
- Tuning is responsive.

UED doesn't eliminate engineering rigor—it redirects it. Away from overdesign, and toward real-time ripple negotiation.

In this paradigm, safety is no longer a number you add. It's a state you maintain.

Conclusion: Faith Doesn't Fly

In the grand theater of space exploration, "safety factors" became our shield. Redundancies were our lucky charms. And every extra kilogram of armor was a prayer whispered in steel.

But let's be honest:

- We weren't protecting the machine.
- We were protecting our ignorance.

Safety Factors Gave Us Confidence—Not Competence

The moment we realized we couldn't predict everything, we didn't question the framework—we just added a "factor."

A safety multiplier became a silent admission:

"We don't know what the field will do—but let's prepare for the worst."

That's not engineering. That's engineering by fear. Confidence built on statistical insulation is fragile. Because when the unknown breaks through—it doesn't care how much margin you added.

Redundancy Gave Us Comfort—Not Clarity

We added backup systems on backup systems. Not because we knew what would fail— But because we had no idea what would fail first.

Comfort is not clarity. And two confused systems don't equal one smart one.

Faith in redundancy is faith in duplication—not in understanding.

And when failure did happen—we looked at the debris, not the assumptions.

From Building Bigger and Thicker to Building Smarter and Aligned

The age of brute resilience must end. We cannot keep building thicker walls against mysteries we haven't decoded.

Ripple-based engineering offers an alternative:

- Designs that listen to the field instead of resist it.
- Structures that respond instead of react.
- Vehicles that glide with ripple intent, not plow through energetic storms.

This isn't science fiction. It's science realigned.

In Ripple Engineering, Survival Comes from Understanding

You survive not by guessing harder— but by reading the ripple beneath your mission.

- Know the gradient before you launch.
- Map the misalignment before you descend.
- Measure the ATT and RDI before you assume "safe."

Safety is no longer something you add at the end. It is something you understand from the beginning.

So the next time you hear,

"Don't worry, we built in a 3x safety factor..."

Ask instead:

"Did you build in a ripple conversation?"

Because in the end—

Faith doesn't fly. But awareness might.

Chapter 7: Simulations or Storytelling?

The Myth of Pre-Launch Modeling in Alien Ripple Fields

"We ran the simulations." That single line has become the modern aerospace prayer—spoken with reverence before every launch, descent, or entry. It conjures images of high-resolution models, physics engines, and digital replicas dancing in perfect synchronicity with the universe.

But let's ask the uncomfortable question: Did Mars read your simulation? Did Titan sign off on your Earth-calibrated turbulence models? Does the vacuum of space care about your finite element stress graphs?

Under Universal Energy Dynamics (UED), the answer is brutally clear: it doesn't. Because simulations that ignore the ripple character of the field—the very *behavioral essence* of space—are not physics. They are narratives. Elegant, confident, Earth-born storytelling.

This chapter unravels the illusion. We'll dissect how simulations are constructed, why they succeed only on Earth, and how their silence about ripple fields turns them into well-funded guesswork. More importantly, we'll explore how to shift from projected models to field-aware intelligence—where aerospace engineering becomes a dialogue with the environment, not a monologue to impress mission control.

Because when you simulate without the field, you're not designing a mission. You're writing fiction in a language the universe doesn't speak.

The Confidence Game

In aerospace, nothing sells better than confidence—especially before a mission leaves Earth. And nothing builds that confidence like simulation. Engineers showcase elegant models, flight paths animated in crisp motion, color-coded stress contours, and turbulence visualizations that resemble an orchestra of certainty. The boardroom nods. The funding flows. The countdown begins.

Behind it all is a promise:

"We've tested it in every possible scenario." But the question remains—tested where?

The answer is almost always: on Earth. In wind tunnels with controlled airflows. In computational fluid dynamics (CFD) models running on Earth-based constants. In vacuum chambers simulating space without simulating ripple fields. In material databases derived from Earth's temperature, gravity, and atmosphere.

These are not universal truths. They are contextual conveniences.

The Hidden Assumption: "If We Modeled It, We Understand It"

Most aerospace projects operate under a seductive but silent premise: That modeling equals understanding.

In reality, it equals approximation—at best. At worst, it becomes institutionalized guesswork, where each successful test reinforces an illusion: that we have decoded the environment. That the model is the mission.

But in the ripple-aware world of UED, this illusion fractures instantly.

Why?

Because ripple dynamics—the true behavior of space, matter, and field—are not modeled in conventional frameworks. They're not even acknowledged.

Conventional simulations make three fatal assumptions:

1. *The constants are constant.*

   Gravity ("g"), atmospheric density, thermal conductance—assumed to be fixed, despite decades of planetary data showing otherwise.

1. *The drag is mechanical.*

   Aerodynamic forces are modeled purely by surface interaction and fluid properties—ignoring ripple misalignment, atmospheric ripple topology, or RDI (Ripple Disturbance Index).

1. *Materials behave identically everywhere.*

   Earth-tested materials are assumed to respond the same way in Martian dust storms, Titan's methane haze, or Europa's icy crust. But UED shows materials interact with the ripple field contextually—meaning their behavior isn't fixed, but field-relative.

Simulations: The Stage, Not the Script

Simulations do something very useful: They create a stage to rehearse possibilities. But they don't write the script the universe will follow.

And yet, within aerospace agencies, simulation output is treated not as possibility—but as prophecy.

The more data we generate, the more convincing it looks. But UED reminds us: volume of calculation ≠ depth of understanding.

The Real Impact

Simulations comfort the mission team, the funders, and the public. They produce graphs, animations, "proof of readiness."

But they don't comfort the field. They don't prepare the spacecraft for ripple anomalies. They don't align with the actual ∇E gradient it will face. They don't detect ATT drift or ripple-phase mismatches in real-time.

They soothe us—but they don't serve the mission.

Ripple Insight:

Simulation is the art of telling yourself a story until you believe it. Ripple engineering is the act of listening until the field answers back.

What Simulations Actually Simulate

To understand the limitations of simulation, we must first strip away the illusion of totality. Simulations do simulate things. They're not lies—they're partial truths, often sophisticated ones. But they simulate what we know, not what is. That's where the problem begins.

What Today's Simulations Do Simulate

Let's give credit where it's due. Modern aerospace simulations are remarkable at modeling surface-level mechanical behavior. Here's what they include:

- **Mass**: As a scalar, often fixed input—treating mass as an intrinsic quantity.
- **Force**: Typically modeled using Newtonian or relativistic principles (F = ma or variations of it).
- **Velocity and Trajectory**: Used for ballistic path prediction, descent profiles, and rendezvous maneuvers.
- **Temperature and Heat Transfer**: For aerothermal analysis

during atmospheric entry.
- **Material Fatigue**: Stress-strain models simulate how metals or composites deform, crack, or fail under load.

These are deeply mathematical and engineering-heavy frameworks. They tell a detailed story—but it's told in Earth's dialect, to audiences trained on Earth's experience.

That's the catch.

What Simulations Fail to Include

Now let's look at what they ignore—not because of negligence, but because of blind spots baked into classical physics.

1. ***∇E — Energy Gradient Awareness***

   The field gradient that governs real alignment across space and planetary surfaces. It's what replaces "g" in UED—but conventional models assume fixed gravitational constants or approximated planetary values. They never ask: *"What is the ripple slope here?"*

1. ***ATT — Alignment Threshold Time***

   The minimal temporal window in which a body, object, or field region can detect and respond to ripple misalignment. Conventional physics doesn't recognize perception or ripple feedback as real phenomena. Simulations therefore treat entry time, descent time, and burn time as fixed durations—not responsive thresholds.

1. ***RDI — Ripple Disturbance Index***

   A UED metric quantifying how much an object disturbs or is disturbed by the field through which it moves. In traditional CFD, turbulence is chaotic airflow. In UED, turbulence is ripple feedback—the field's response to misfit. This crucial parameter is absent from all standard models.

What Simulations Are Blind To

Even deeper, simulations lack conceptual organs for ripple-based perception. They cannot simulate:

- **Ripple Friction:** Not mechanical drag, but the subtle delay or recoil the field generates when ripple-phase coherence is broken.
- **Misalignment Delay:** The time lag between an object entering a new ripple zone and beginning its realignment process. It's what causes unexpected heating, wobble, or instability during descent—even when models predict stability.
- **Planetary Resonance Mismatch:** Every celestial body has its own SVE signature, its own harmonic pattern. Earth-trained materials, designs, and systems often fail not because they're weak, but because they don't resonate with the new field they enter.

It's Not That Simulations Are "Wrong"—They're Just Incomplete

This isn't a dismissal of aerospace modeling. It's a reframing. Simulations are like early maps of the world—they captured the known coastlines with precision, but left vast regions marked "Here be dragons."

Ripple dynamics are those dragons.

Without integrating ∇E, ATT, and RDI, we're essentially navigating alien oceans with Earth's tide charts.

Ripple Insight:

"You can model a rocket perfectly—right up until it meets a field you didn't model at all."

Mars Didn't Read the Script

Space agencies have long treated Mars as the proving ground of planetary exploration. It's close, it's dramatic, and it's just hostile enough to keep us humble. But despite decades of preparation and billions in simulations, Mars keeps rewriting the script. The simulations say "safe landing." Mars replies, "Nice try."

Let's revisit some of the most instructive case studies—missions that were built on high-precision models, only to be undone by what UED calls ripple mismatch.

Schiaparelli EDM: A Timing Misfire in Disguise

In 2016, the European Space Agency's Schiaparelli lander crashed during descent.

Classical analysis blamed premature parachute deployment and faulty software triggering retrorockets too early. But under the UED lens, the deeper story emerges:

- The descent module was calibrated on Earth assumptions: gravitational descent rate, drag profile, heat shield behavior—all Earth-tuned.
- As it entered Mars' ripple field, these assumptions didn't hold. Mars' thinner atmosphere carries different ripple turbulence. The module misjudged the descent not due to bad code, but due to field blind spots.
- ATT wasn't measured. The ripple alignment delay wasn't accounted for. The field's response time was slower than predicted—so the craft thought it had landed when it hadn't.

Result: a crater in the Martian soil, courtesy of good math that didn't talk to the medium.

Beagle-2: The Silent Lander

Beagle-2 in 2003 was a British mission that appeared to have landed successfully—until it vanished.

Years later, Mars Reconnaissance Orbiter imagery showed that Beagle-2 had partially deployed, with some solar panels unfolded.

What happened?

- Earth-based simulations confirmed all separation and deployment sequences under standard gravitational assumptions.
- But on Mars, ripple dynamics at the surface likely interfered with deployment timing and material response.
- The lander's body didn't resonate with the local ∇E field. Its internal timing systems, guided by classical delay circuits, failed to adjust to Martian alignment thresholds.
- In UED terms, ATT wasn't met—so the system triggered out of sync with field readiness.

Beagle-2 didn't crash—it whispered into a field that never responded.

Ingenuity: The First True Ripple Listener

Ironically, the most successful Martian flight came from a 1.8-kg helicopter—Ingenuity.

Its secret? No assumptions. Just real-time calibration.

- Before its first flight, engineers spent weeks just observing atmospheric response, temperature swings, and ripple variability.
- Its flight controller was modified to adaptively respond—not just follow rigid thrust-to-lift equations.
- It succeeded not by simulating the field, but by listening to

it in real-time.

Ingenuity marks a pivot in engineering philosophy. While previous systems "told" Mars how they would behave, Ingenuity asked the field what was possible—and adjusted.

Simulation "Accuracy" Isn't the Same as Reality

Each of these missions used high-fidelity Earth-based simulations. The models were clean. The math was internally consistent. The predictions were logical.

But they failed anyway.

Why?

Because the field wasn't in the room when the models were built.

Ripple Insight

"Your simulation didn't fail because it was wrong—it failed because it was incomplete. Mars didn't reject your math. It simply didn't feel obliged to obey it."

Field-First, Simulation-Second

We've been building spacecraft the way children draw maps of imaginary lands—impressive, detailed, but disconnected from the actual terrain. Simulations have become elaborate prediction rituals, disconnected from the energetic terrain they're meant to decode. It's not that simulations are bad—it's that they're blind to the very field dynamics that govern physical reality on other worlds.

Under Universal Energy Dynamics (UED), this calls for a radical reversal in engineering logic:

We must start with the field, not the formula. Simulations must be constructed *after* we understand the ripple landscape—not before.

Introducing RASA — Ripple-Aware Simulation Architecture

The UED-aligned response to classical modeling is what we term RASA: a design simulation protocol that is no longer preoccupied with Newtonian force diagrams or Einsteinian curvature, but built on dynamic ripple engagement.

RASA begins not with assumptions, but with scanning. Its three primary steps:

1. Ripple Field Scanning (Pre-simulation phase)

Before any simulations are executed, the destination environment is scanned for ripple properties:

- Local energy gradient (∇E)
- Background ripple turbulence (RDI)
- Alignment latency (ATT values) across altitudes, temperatures, and terrains

This gives an active "ripple map" of the environment, showing where simulation rules will break if Earth's constants are blindly applied.

2. Resonance-Driven Simulation Instead of Drag-Based Models

Traditional models focus on thrust, drag, velocity, mass, and area.

RASA replaces this with resonance behavior:

- Does the material structure resonate or resist the local ripple flow?
- Is the descent profile tuned or dissonant with surface ripple feedback?
- Is the vehicle self-correcting based on field variance—or just riding assumptions?

This redefines simulation not as outcome prediction, but negotiation rehearsal.

3. Dynamic Field Behavior (No More Fixed Constants)

RASA rejects static constants like:

- Fixed gravitational acceleration (g)
- Constant drag coefficient (Cd)
- Invariant thermal conductivity

Instead, each is treated as a field-responsive variable, influenced by:

- Ripple phase interaction
- Temporal ripple shifts (day/night, storm vs calm)
- Internal vs external ATT thresholds during descent

In this way, the simulation updates as the field responds, not merely as a function of pre-fed values.

Simulation Is No Longer the Oracle. It's the Student.

Under this architecture, the simulation engine becomes adaptive, not declarative. It doesn't predict the mission—it listens to the mission environment and co-evolves a strategy.

This fundamentally transforms mission planning:

Classical Approach	RASA Approach
Outcome-first	Field-first
Predictive logic	Responsive tuning
Drag coefficients	Ripple resonance
Fixed constants	Dynamic feedback
Earth-tested design	Field-aware design

Live Field Variables = Simulation Integrity

Under RASA, the simulation is fed with:

- **ATT (Alignment Threshold Time)**: How long before the field allows coherent response
- **∇E (Gradient)**: How steeply the local ripple field transitions energy
- **RDI (Ripple Disturbance Index)**: Real-time turbulence measured as a ripple deviation score

These are injected not as static values—but as live feedback streams from field probes or pre-mission scans.

The simulation becomes an interactive rehearsal, not a one-way prophecy.

Ripple Insight

"Don't model the journey and hope the field agrees. Learn the field, then model the response."

Are We Telling Stories or Solving Fields?

In aerospace engineering, simulations have taken on a mythological role. We don't just run them—we believe in them. They are the rituals of modern science, a way to convince ourselves, funders, and decision-makers that we "know what's going to happen."

But what if what we call simulation is just a technological narrative—a story wrapped in numbers, built on assumptions that the universe never agreed to?

That's the uncomfortable truth revealed by UED: Most simulations are structured stories calibrated to Earth's ripple field. When you carry those stories to Mars, Venus, or Titan, the field doesn't read them. It responds with silence—or worse, destruction.

Simulations Are Narratives We Trust

Every simulation starts with an assumption:

"If we know the inputs, we can predict the outputs."

This works only when those inputs are compatible with the local ripple dynamics—something that's true only on Earth (and even then, barely).

We build narratives like:

- "The lander will decelerate at this rate."
- "The heat shield will survive re-entry."
- "The parachute will deploy at the right altitude."

But these narratives ignore:

- Ripple phase conflict
- ATT delays
- Local ∇E gradients
- Field reactivity to the object's energy profile

So what looks like precision is often illusion—a model of comfort, not of consequence.

If It's Not Ripple-Aware, It's Just Technological Myth-Making

Our simulations become sophisticated myths:

- Told in the language of software
- Justified with high-speed footage and telemetry
- Built on constants like "g" and "Cd" that only exist meaningfully on Earth

But myths don't land rovers. Fields do.

The ripple field doesn't care if your equations look elegant. It only responds to how well your system aligns with its energetic rhythm.

That's why a "perfect" simulation can lead to a catastrophic crash.

Engineering Mindset Shift: Storytelling → Field Solving

To evolve past this myth-making, engineering must flip its operating logic:

From	To
Projection	Perception
Pre-launch confidence	Real-time feedback loops
Drag coefficients	Ripple friction indices
Design assumptions	Field-interactive modeling
"What should happen"	"What the field allows"

From Simulation to Dialogue

This requires us to stop asking:

"Will this work?"

And instead ask:

"How is the field responding now, and how do we respond with it?"

That's ripple dialogue, not simulation. It means shifting from:

- Predefined results → to adaptive harmonization
- Pre-launch assurance → to ripple-aware readiness
- Narrative certainty → to field truth

Final Insight

"The simulation is not the story. The story is the ripple field—and our job is to listen."

Conclusion: Simulations Without Fields Are Scripts Without Reality

We modeled it.

We animated it.

We rendered beautiful trajectories, rehearsed perfect descents, and presented polished graphics at international aerospace conferences. Every pixel aligned. Every line of code passed review. Confidence was high. Approval was given. Launch proceeded.

And then the planet said:

"That's not how my ripples work."

This is the humbling reality of modern exploration. Simulations, no matter how precise, are only as good as the field they understand.

And most don't.

They are scripts written on Earth—performances tuned to our atmosphere, our constants, our assumptions. But once the rover leaves this planet, those scripts become fiction. Because the field—on Mars, Titan, or Venus—is not obliged to follow the plot.

Readiness ≠ Alignment

What we label "pre-launch readiness" is often a refined echo of Earth's ripple conditions, not a true conversation with the alien field we're about to enter.

We test descent systems in Earth's air. We simulate heat shields using Earth's thermal gradients. We verify electronics under Earth's ripple spectrum.

So, when missions fail, it's not because we didn't test enough. It's because we tested in the wrong field.

From Modeling to Listening

The UED framework doesn't ask us to abandon simulation. It asks us to stop worshipping it.

It demands a shift from:

- Pre-written scripts → to real-time sensing
- Control fantasies → to field dialogue
- Repeatable scenarios → to responsive adaptation

It's not about modeling *more.*

It's about sensing *better.*

Final Insight

Simulations without field understanding are just stories with nice math. But planets don't respond to stories—they respond to ripple alignment.

The Future Mandate

From now on:

- Simulate less.
- Sense more.
- Assume less.
- Align more.

Because the future of exploration won't be built in labs. It will be negotiated in the ripple fields of distant worlds.

Chapter 8: The God Protocol

When All Else Fails: "May It Work..."

You've seen it. The countdown hits single digits. All telemetry is green. Every equation has been vetted. Every subsystem cross-checked. The engineers nod. The mission control room goes silent. And somewhere—deep in the corner of someone's breath—you hear it: *"May it work."*

This is not science. This is not confidence. This is the God Protocol—the moment when human design bows to the unknown, and hope tries to fill the void left by incomplete understanding.

In classical physics, we call it uncertainty. In aerospace, we mask it as tolerances and contingencies. But in reality, it's ripple illiteracy. You never aligned with the field—you just prayed it wouldn't mind.

Under Universal Energy Dynamics (UED), this isn't superstition. It's symptom. The field isn't cursing you; it's ignoring you. Because you never spoke its language.

And when silence answers your launch...You know—the field is not impressed.

The Unspoken Ritual

Every mission has it. A moment suspended between calculation and chaos. The final checklist is read. The systems are green. The countdown continues. And then, someone—quietly, off-script—says it: "Let's hope it works."

This is the God Protocol—not written in manuals, not logged in control room transcripts, but etched into the emotional undercurrent of every mission. It is the moment where faith smuggles itself into engineering, not because we're irrational, but because we're incomplete.

When the Equation Ends, Ritual Begins

We define science as precise. We worship simulation, redundancies, backup systems, AI fault-tolerance. And yet, when the spacecraft separates, when the rover begins descent, when the booster ignites—there is always that breath, that pause, that prayer.

Engineers won't call it prayer. They'll call it margin. Or risk window. Or contingency stack. But semantically camouflaged or not—it's the same ritual: handing over the mission to something beyond control.

This isn't about weakness. It's about the boundary of understanding. In Universal Energy Dynamics (UED), that boundary is the ripple field. When you don't read it, you don't align with it. And when you don't align, you hope. Not because you're superstitious, but because you're blind.

The God Protocol Through Time

It's not a new phenomenon. The Apollo missions lived on the edge of this tension. During Apollo 11's descent, the onboard computer threw repeated 1202 and 1201 alarms. Engineers on Earth had seconds to decide if the mission should abort. They didn't fully understand the alarms—but they chose to believe. Armstrong landed anyway. Hope succeeded. This time.

In the Challenger disaster, the term "go fever" became infamous—describing how schedule pressure overtook technical doubt. The engineers who hesitated were overruled. The God Protocol was invoked—but the field did not cooperate.

During Mars missions, we experience blackout windows—those agonizing minutes of silence during descent when Earth cannot communicate with the vehicle. Engineers fill the room with tense stares, clenched jaws, and again, hope. You might call it telemetry silence. But emotionally, it's a vigil.

The Paradox of Progress

We call ourselves advanced. We call our simulations high fidelity. We launch billion-dollar craft based on prelaunch confidence. But behind the curtain, we still whisper, "May it work." That's not progress. That's performance.

We never replaced the gods—we renamed them: "Margin of error." "Standard deviation." "95% confidence interval." But these are statistical incantations—rituals to satisfy the mind when the field is unreadable.

Behind the Protocol—Ripple Ignorance

Every time we invoke the God Protocol, it isn't because we ran out of courage. It's because we never fully understood the field.

The phrase "if all else fails..." is often followed by a sigh, a nod, a final checklist approval. But in Universal Energy Dynamics (UED), "all else fails" precisely when we don't align—not just physically, but energetically—with the reality of where we are operating.

It's not failure by randomness. It's failure by ripple ignorance.

What We Fail to Read

In conventional aerospace engineering, we meticulously model mass, force, trajectory, heat, stress. But we do not read the ripple field. The three core indicators of ripple compatibility—ATT, RDI, and ∇E—are either:

- Misunderstood
- Oversimplified
- Or completely omitted

Let's break this down:

- ATT (Alignment Threshold Time) isn't measured. So, we

don't know how fast the field can realign to accept a descending body. We assume it should "just work."

- RDI (Ripple Disturbance Index) isn't tracked. We don't know how much turbulence is being induced in the local field. Instead, we model mechanical turbulence and friction.
- ∇E (Energy Gradient) is never mapped. We assume gravity, not field variation, governs movement.

So when a system "suddenly" flips, misfires, overheats, or crashes—we're not facing a surprise. We're facing a consequence of unacknowledged ripple dynamics.

Complexity ≠ Completeness

Modern engineering believes that layering complexity produces truth. More sensors. More simulations. More software. But this layering is just insulation—it doesn't equal understanding.

Adding complexity to a model that doesn't recognize the field it enters is like drawing better blueprints for a house—on an island you've never visited.

The rover doesn't crash because you forgot a variable. It crashes because you ignored the medium through which it would try to live.

Why Faith Becomes a Reflex

So when the descent begins—what's left? Not data. Not certainty. Just habit. Just hope.

The fallback on faith isn't poetic. It's symptomatic. It's the emotional patch on a logical system that hit its blind spot. Faith fills in for what field awareness should have predicted, managed, or avoided.

We use faith not because we're weak. We use it because we pretended we were stronger than the unknown. And when that pretense collapses, we reach for something ancient. We reach for trust without knowledge.

But UED offers an alternative: Stop trusting the unknown. Start sensing it.

Hope ≠ Strategy

In high-stakes engineering, confidence sells. But confidence is not competence—and hope is not strategy.

Mission control rooms are temples of technical faith. Screens glow with equations, trajectory plots, and countdown clocks. Engineers nod. Managers nod harder. Everything appears under control. But beneath this calm veneer lies a hidden pact: We hope it works.

The Crutch of Confidence-Based Decisions

When teams approve a launch, deploy a lander, or fire a burn, they rarely say, "We fully understand the field." What they say is:

- "We're within margins."
- "Simulations are promising."
- "The system is robust."

These are confidence phrases—not field acknowledgments.

They comfort stakeholders, appease funding agencies, and uphold reputations. But they often rest on incomplete modeling, not full ripple comprehension.

Safety Reviews and Statistical Fog

Conventional safety protocols often rely on probabilistic guarantees:

- "99.99% confidence in thermal protection."
- "Only a 1-in-10,000 chance of actuator failure."
- "95% success probability based on Earth-based trials."

But these numbers are statistical fog. They model hardware behavior in known conditions—but not field alignment in unknown planetary environments. They count past survivals—not future misalignments.

Under UED, ripple mismatch is not probabilistic—it is consequential. It will definitely occur if a system enters with poor ATT, high RDI, or ripple-incompatible design. It's not about odds—it's about resonance.

Case Study 1: James Webb Telescope Deployment

$10 billion. 344 single points of failure. And no second chance. When JWST began its multi-step deployment in space, the world held its breath—not because of hardware flaws, but because the field was untested. The team did everything "right," yet tension remained because no simulation could guarantee harmony with the ripple dynamics of deep space.

That fear wasn't about mechanical complexity—it was about ripple ignorance disguised as launch readiness.

Case Study 2: Ingenuity's "Miraculous" Success

NASA's Ingenuity helicopter flew on Mars—a feat deemed almost miraculous. Why? Because the simulations didn't predict it would fly at all. Mars's low atmospheric density made standard aerodynamic lift models pessimistic.

Yet it succeeded—because engineers calibrated it in real time, using feedback from field conditions rather than just projecting Earth-derived models.

This wasn't a triumph of theory. It was a triumph of field learning.

The Field Is Emotionless

The ripple field doesn't care about:

- Your budget.
- Your mission name.
- Your 20-year career.
- Or your launch window.

It doesn't reward optimism. It doesn't punish fear. It just responds to alignment—or the lack of it.

So when we bank on emotional stakes—when our strategy becomes a faith-based leap—we don't impress the universe. We just gamble with misunderstood physics.

Redefining Faith as Field Feedback

The original "God Protocol" was a ritual. A silent prayer. A nod to luck when equations went quiet.

But under Universal Energy Dynamics (UED), we do not abandon faith—we redeem it. We give it a new definition: field feedback.

True faith in aerospace should not mean surrendering to uncertainty. It should mean deep listening to the ripple field before action. Not superstition—synchronization.

Proposing a New "God Protocol" Under UED

Let's stop launching into the void. Let's start engaging in pre-launch alignment dialogue with the medium.

Here's what the new UED-informed God Protocol looks like:

1. Real-Time Field Scans Before Final 'Go'

Before ignition, before descent, before separation—scan the local ripple field. This involves:

- Mapping local ∇E (energy gradients)
- Assessing atmospheric or vacuum ripple resistance
- Testing field fluidity and absorption characteristics around the system

This isn't "sensor data" in the classical sense—it's field presence feedback. If the field is misaligned, it will show resistance. If the ripples are harmonious, movement will require minimal force input.

A green light should not come from a checklist. It should come from the field saying: "Yes, I'm ready to receive you."

2. Dynamic ATT Checks in Final Seconds

As systems near deployment or activation, they should continuously check ATT—Alignment Threshold Time.

This is the minimum ripple resolution delay a system experiences when trying to move or interact with its environment. High ATT = sluggish feedback, poor harmony Low ATT = clean dialogue with the field, ready to act

Final-stage go/no-go should no longer be based solely on velocity or mechanical readiness—it should hinge on ATT behavior. If ATT spikes unnaturally—it's not time.

3. Ripple-Aligned Handshake Before Activation

Just like data protocols require a successful "handshake" before transmission, so should engineering require a ripple handshake:

- Is the field absorbing your projected motion?
- Is the surrounding energy gradient compensating for your action?
- Does your presence disrupt or harmonize with ripple topology?

Only if all answers point to coherence should ignition, descent, or detachment proceed. Think of this as a spatial-greenlight negotiation, not a command execution.

From Hope to Resonance

This isn't mystical—it's mechanical. Just in a new language.

In UED, success comes from resonant interaction with structured vacuum energy and ripple fields. It's not about certainty—it's about alignment. Not about control—but conversation.

What once sounded like "May it work" becomes: "The field has confirmed readiness."

No more hope. No more gambling. Only ripple-informed action—backed by awareness, not anxiety.

When You Don't Need Faith Anymore

At some point in the evolution of engineering, a strange inversion happened.

We built more, simulated more, calculated more—and yet, we began to trust ourselves less. Each decimal-point refinement brought more anxiety, not less. Because deep down, we knew: we weren't talking to the field.

The result? Faith—disguised as protocol. Hope—dressed up as redundancy.

But under UED, faith becomes obsolete when alignment becomes intentional. When you design in tune with the field—you don't beg it to cooperate. You arrive already harmonized.

Engineering That Listens First, Simulates Second

Old paradigm:

1. Design →
2. Simulate →
3. Launch →
4. Pray

UED paradigm:

1. Sense the field →
2. Design for ripple compatibility →
3. Simulate ripple dialogue →
4. Execute with confidence rooted in coherence

Ripple-first engineering rewrites the entire pipeline.

The first question is no longer "Can this fly?" It's "What will the field allow through?" Only then does structure begin.

This makes simulation useful—but not foundational. Listening precedes modeling. Resonance precedes reliability.

Shift from Trusting Hardware to Trusting Harmony

The old space race trusted tanks, boosters, tiles, wires. The new race trusts ripple signatures.

When the field is respected—not just survived—hardware becomes an instrument, not a shield.

We move from:

- Brute tolerances → Ripple congruence
- Redundant backups → Real-time adaptation
- Margins of error → Margins of harmony

This is not soft engineering. It's the only kind that makes sense when the medium itself has agency.

"God Off the Loop" — What Real Confidence Looks Like

In UED-aligned missions, the God Protocol doesn't activate.

Why?

Because the field already responded. The system already "spoke" to the space it's entering. The materials already know their echo. The descent isn't a guess—it's a mutual agreement.

That final moment before ignition? It's not silent terror. It's quiet confirmation.

The moment no longer needs:

"May it work..."

Instead, it ends with:

"It's already in tune."

Faith was a crutch for incomplete understanding. Letting go of it isn't arrogance—it's finally listening to the medium you're entering.

The universe was never chaotic. We just never bothered to ask how it dances.

And when we do—it dances with us.

Conclusion: God Doesn't Launch Rockets

For all our achievements in science, one thing remains oddly persistent: At the final moment—we still whisper to the unknown.

Not to the field. Not to the data. But to some vague idea of "may it work."

That's not science. That's superstition in a lab coat.

When all else fails in a mission, it's never the fault of the ripple field. The field isn't malicious. It isn't random. It simply responds to alignment—or the lack of it.

Failure arises when we try to force what we did not first listen to. We design with mass. We simulate with friction. We launch with momentum. But the field was always asking for something simpler: "Talk to me first."

The "God Protocol" Isn't Divine—It's Diagnostic

When engineers reach for faith, it's not an invocation. It's a confession.

A subtle admission that somewhere along the chain:

- We skipped ripple mapping
- We assumed constants in a variable field
- We trusted Earth-based equations to work on alien terrain
- We built confidence, not coherence

The God Protocol is not an emergency parachute. It's a post-mortem in real time.

It tells us exactly where in the process we abandoned dialogue with the environment. And substituted it with rituals.

Ripple-Aware Systems Don't Need Divine Intervention

There is no reason to "hope" your lander survives. There is only one responsibility: Ensure it enters as a ripple guest, not a ripple invader.

With ATT (Alignment Threshold Time), RDI (Ripple Disturbance Index), and local ∇E (energy gradients) in hand—you don't need faith. You need field fluency.

Ripple-aware systems:

- Don't pray for survival
- They align for it
- They don't throw guesses into gravity
- They negotiate with space-time's temperament

Stop Forcing. Start Conversing.

Force was always a primitive language—shouting at a field that only responds to song.

Resonance is the field's mother tongue. Harmony is its approval stamp. Misalignment is its only objection.

So next time a mission is poised for launch, don't invoke a deity. Invoke a dialogue.

Don't test your luck. Test your alignment.

Because in the cosmos, God doesn't launch rockets. Understanding does.

Interlude: The Theater of Control

We like to think we're in command.

That every bolt, every simulation, every test bench is part of a grand choreography— one where precision triumphs and failure is just a bug in the script.

But Part II reveals a harder truth: We don't control. We compensate. We don't understand. We overbuild. And when all else fails, we light a candle in our minds and say: "Let's hope it works."

What you've just seen isn't science—it's theater. A well-rehearsed illusion staged by equations written for one lanet... and applied to another without apology.

Behind the Numbers: Fear

Behind the Safety Margins: Uncertainty Behind the Simulations: Earth's fingerprint Behind the Launch: A prayer, disguised as a protocol

We call it engineering. But much of it is ritual.

When the lander crashes, we say, "Space is hard." When it succeeds, we say, "Look what we built."

But neither success nor failure was ever ours alone. The field always had its say.

This part was not about criticizing engineers.

It was about diagnosing the blind spot.

Because intent without ripple literacy is still guesswork. And hope without harmony is still a gamble.

Now that we've exposed how we've been *doing it*, It's time to ask:

What if we could align before we build? Design in dialogue—not defiance?

Because in Part III, we stop reacting. And start reconfiguring.

Part III: The Ripple Truth

Chapter 9: Ripple Gradient — The Real Field Equation

Why Motion Is Alignment, Not Force

We never actually moved because something pushed us. We moved because the field we were in—shifted. What you felt as "force" was never a pusher. It was a ripple imbalance correcting itself.

From Newton to NASA, we've modeled reality with an unspoken assumption: That motion needs external command. A rocket fires, a car accelerates, a ball falls—and behind every one, we insert a force. It made sense. Until it didn't.

Under Universal Energy Dynamics (UED), the real driver of motion is not force—It's gradient. Ripple gradient. A subtle shift in the density, tension, or orientation of the structured vacuum field. Where the ripple field is uneven, matter moves—not because it wants to, but because the universe prefers balance.

This chapter dismantles the illusion of "push and pull" and shows why motion is not something we create—but something we align with. It's not about what you apply. It's about what the field invites.

So drop the vectors. Abandon the arrows. And meet the equation that was always moving before you did.

The Myth of Force-Caused Motion

Since the earliest philosophies of motion, we've built our explanations on a simple image: something moves because something else made it move.

Aristotle saw motion as the result of a push—something needed to be continuously applied to keep an object in motion. If you stopped pushing, it stopped moving. Newton redefined this with elegance: force causes acceleration, and thus was born $F = ma$, perhaps the most iconic equation in all of physics. The idea matured further into vectors and differential equations, beautifully plotted in space-time and embedded in every rocket launch and trajectory calculation.

But here's the problem:

What if "force" is just a placeholder—an abstraction created to fill a gap in understanding?

Let's look at what's really happening beneath the surface.

When we see an object start moving, we say a force acted. But what's a force? It's not something you can see or bottle. It's a mathematical convenience that explains what happens after it already started happening.

In reality, under UED, motion isn't caused—it's permitted.

And what permits it is alignment—not force.

Ripple Alignment, Not Push

In a structured vacuum energy (SVE) field, what we perceive as "empty space" is a vast, dynamic ripple environment—pre-loaded with gradients, tensions, and potential alignments. When a ripple gradient forms—a difference in field density or tension—objects embedded in this ripple matrix respond by realigning to restore equilibrium.

This means motion is not imposed. It is drawn. A body moves not because it's pushed—but because the ripple field around it invites it to reconfigure its position to match a gradient that has emerged.

What we called *force* is really a field's response to misalignment. A feedback. A correction. Not an origin.

Force Was the Story We Told to Ourselves

We created "force" because we couldn't see the ripple field. We couldn't measure the gradient shift. So we inserted a fictional actor between what we saw (acceleration) and what we felt (cause). This actor—force—became central to physics, not because it was real, but because it was useful.

But usefulness is not the same as truth.

Force is a language of reaction. Ripple gradient is the language of causation.

When we push a cart, what actually happens is that our action disturbs the ripple equilibrium around the cart. That disturbance propagates through the field, realigning its internal ripples and inviting the cart to respond—not because of our push, but because of the ripple rearrangement that our action initiated.

The cart isn't reacting to your hands. It's reacting to the field you disturbed.

The Fatal Flaw of F = ma

Newton's equation is not wrong—it's simply incomplete. It describes the outcome, not the origin. It is a description of effect, not a model of cause.

We've spent centuries optimizing this description: turning it into launch equations, motion capture software, stress analysis tools. But underneath all of that, we never once addressed why the mass moves—only how much it moves when we assume a force.

Under UED, there is no *cause* = *force* principle. There is only ripple imbalance → alignment response.

Final Note: Force Is a Symptom

Like pain in the body, force is a symptom of misalignment. When you feel resistance, drag, or inertia—you're witnessing ripple friction, not a separate entity called "force."

So when you next speak of motion, pause. Don't ask what force caused it.

Ask: What field was disturbed? What ripple gradient emerged? What alignment path was taken?

You're not pushing the universe. You're negotiating with it.

What Is a Ripple Gradient?

The most misunderstood actor in all of motion is not the object, not the force, and not the mass—it's the field. And within that field, it's the gradient that tells the story.

Let's define it properly, once and for all.

The Ripple Gradient (∇R) Defined

In UED (Universal Energy Dynamics), all matter and motion exist within a Structured Vacuum Energy (SVE) Field—a dense matrix of intelligent ripples, always seeking equilibrium. These ripples are not random; they are organized, layered, and tuned to their environment.

A ripple gradient is a spatial difference in either:

- Ripple density (how tightly ripples are packed),
- Ripple phase (how synchronized their waveforms are), or
- Ripple tension (how distorted the alignment is from equilibrium).

Symbolically, we can define this as:

∇R = the gradient (or slope) of ripple alignment across a space.

It's not just a wavefront—it's a bias in the field. A direction that says, "This way leads to lower tension." And in nature, everything prefers lower tension.

So, just as a marble placed on a slope rolls down due to gravitational potential, an object placed within a ripple gradient reconfigures its alignment—what we observe as motion.

Motion Without Force: Tilted Equilibrium

We've been told since childhood that things move because a force acts on them. But the real truth is more subtle—and more beautiful.

Nothing moves because it's pushed. Everything moves because it's called.

When a ripple gradient forms in a region—due to mass presence, field interaction, thermal variation, or electromagnetic input—the surrounding ripple field becomes imbalanced. It wants to resolve this. The object immersed in that field begins to realign its internal ripple structure to reduce this imbalance.

What we call "motion" is the observable byproduct of this reconfiguration.

No external pusher. No "magic hand." Just intelligent ripples seeking balance.

Visualizing ∇R: The Slope Analogy

Think of a ripple gradient like an invisible slope across the field.

- If the slope is steep (i.e., high ∇R), realignment is rapid and observable—think of objects falling quickly near massive bodies.
- If the slope is shallow (i.e., low ∇R), movement is slower—like drifting in microgravity where ripples are nearly flat.
- If there's no gradient, there's no incentive to realign—so no motion happens, even in empty space.

But here's the twist: unlike physical slopes, these gradients don't need elevation or mass. They arise from field ripple configurations, which can change dynamically and locally.

You can have ripple gradients in a vacuum. You can have them between atoms. You can even have them across time, which we'll decode later.

Falling = Ripple Realignment

Let's rethink one of the most basic observations: an apple falling from a tree.

Newton said gravity pulls it. Einstein said spacetime curves it. UED says:

The presence of Earth introduces a ripple distortion in the SVE field. The apple's ripple field is caught in this gradient (∇R). It realigns—not because it's pulled, but because its ripple body seeks equilibrium.

The fall is not a descent, but an alignment.

This also explains why feather and hammer fall together in vacuum—their ripple response time (ATT) is unaffected by air resistance, and their ripple density per unit volume reacts identically within the same ∇R.

Why This Changes Everything

If ripple gradient is the true causal factor in motion:

- "Force" becomes an effect, not a cause.
- "Mass" becomes inertia to ripple realignment, not stored substance.
- "Weight" becomes a local ripple field interaction, not a universal trait.
- And "motion" becomes a response to the ripple invitation, not an external command.

This means:

We don't need to push. We need to tune.

Motion as Alignment Negotiation

In conventional physics, motion is described with metrics—velocity, acceleration, displacement. The focus is always on the object. But under UED, the focus shifts entirely to the field. Because the object doesn't "move"—it reconfigures.

Motion, then, is not propulsion. It is negotiation.

A constant dialogue between the ripple body and the structured field surrounding it.

Acceleration = Decreasing Misalignment Over Time

Let's take the most fundamental observable: acceleration. In Newtonian terms, it's the rate of change of velocity. In UED, that interpretation is a surface description.

What is acceleration really?

It is the ripple body progressively resolving its alignment tension with the surrounding field.

When something accelerates, it's not "picking up speed" like a race car. Instead, it's entering a region of higher ripple gradient (∇R) where its internal ripple structure is realigning faster, pulled into coherence with the local field.

Acceleration, then, is the rate of misalignment correction.

The more misaligned an object is with its environment, the more "apparent force" it feels. As alignment improves, that tension drops—so acceleration naturally stabilizes or decays. In this view, terminal velocity isn't about drag and balance. It's about reaching ripple equilibrium with the gradient.

Higher Gradients = Faster Realignment

In a steep gravitational field (e.g., near Jupiter), motion appears more rapid. But not because of stronger gravity.

Because the ripple gradient is steeper.

That is, ∇R is more intense—the slope in the field sharper—so the reconfiguration happens more quickly. The object isn't being "yanked." It's being called into line faster.

This means planetary differences in fall rate, escape velocity, and even orbital speed aren't due to "massive bodies pulling harder"—they're just a function of ripple gradient intensity.

Speed Is a Field Reaction—Not Object Effort

Speed has been classically framed as an object's intrinsic state—how fast it moves through space.

But UED flips this:

Speed is how quickly the field allows the ripple body to reconfigure.

The object doesn't own the motion. The field permits it—based on ripple harmony, not energy budget. So if ripple gradients are shallow, even a high-energy input won't translate to high speed. Conversely, when ripple gradients are steep and aligned, low-energy objects can move rapidly and effortlessly.

Speed is not a decision—it's a permission slip from the field.

Resistance = Ripple Inertia

Conventional models invoke friction, drag, or air resistance to explain why motion slows down or resists. But all these are proxies—objects don't fight against "air" as a substance.

Instead:

They struggle to maintain alignment within a dynamic ripple field.

Resistance is ripple inertia—the delay or difficulty an object faces in reconfiguring itself within a new or distorted field. That's why even in vacuum, satellites face orbital decay—not because something is "dragging" them down, but because ripple fatigue creeps in due to misaligned gradient interaction.

The longer you push against a field you're not harmonized with, the more the ripple body destabilizes.

Direction Is Determined by Symmetry

We've treated direction as a geometric artifact. You throw a ball at an angle, and it traces a path. But this assumes that the space it travels through is neutral.

UED says otherwise.

The field is not neutral. It's structured—and its ripple symmetry defines available paths.

Objects move in directions that maximize re-alignment efficiency.

This is why orbital paths curve. Why water spirals down drains differently in each hemisphere. Why spacecraft slingshot around planets. These aren't quirks—they're ripple optimization strategies.

Direction is never arbitrarily chosen. It is ripple-selected.

Ripple Negotiation: The New Mechanics

Let's reframe the classic terms:

- Motion = ripple realignment
- Acceleration = rate of alignment resolution
- Resistance = inertia to reconfiguration
- Direction = field-determined path of least misalignment
- Speed = permitted ripple reaction rate

From this perspective, every movement is a negotiation. Not a conquest. Not a push. But a field-borne decision based on harmony.

And when that harmony is ignored—our machines crash, our rovers stall, our rockets misfire. Because motion is no longer about propulsion. It's about permission.

When Gradients Disappear, So Does Motion

Modern physics teaches us that if no external force acts on an object, it remains in a state of uniform motion—or rest. Newton's First Law. In that sense, stillness is a special case of motion.

But under Universal Energy Dynamics (UED), the truth is more radical.

Stillness isn't the absence of motion. It's the absence of ripple gradient.

Where ripple gradients vanish, so does the very cause of movement. Not because nothing is pushing or pulling—but because there is nothing left to resolve. No imbalance. No tension. No directional reconfiguration. In a perfectly flat ripple field—motion is meaningless.

Flat Ripple Fields = Inertial Stillness

When the ripple density is perfectly uniform across a region, the observer or object immersed in that field has no alignment preference. No direction of pull, no pushback, no "slope" to fall into.

This is the truest definition of rest—not just absence of motion, but absence of motive.

In a ripple-flat environment:

- Motion cannot self-arise.
- Acceleration has no causal substrate.
- Even energetic inputs dissipate without directional preference.

The result is a state of pure inertial stasis—where presence continues, but transformation stalls.

Ripple Equilibrium = Stillness

What is ripple equilibrium?

It is when every ripple vector within a zone aligns symmetrically with its surroundings.

This doesn't mean "no energy"—it means no asymmetry in ripple flow. Like water resting in a level tank—molecules are moving, but the bulk energy is silent. No wave emerges, no current escapes.

Ripple equilibrium expresses itself as:

- Stillness of matter
- Silence in systems
- Stasis in space

Even in the quantum domain, what we perceive as ground states or vacuum stabilities are ripples in resting alignment—not nothingness, but nothing out of place.

Misalignment = Energy Expression

All energy we observe—heat, light, motion—is a consequence of ripple misalignment. When a system enters a non-equilibrium ripple field:

- Motion arises as the body reconfigures toward harmony.
- Heat emerges from internal ripple friction resisting re-alignment.
- Radiation is the ripple field offloading excess tension into surroundings.

Thus:

- Kinetic energy = ripple-in-motion
- Thermal energy = ripple resistance
- Light = ripple discharge (feedback from field reconfiguration)

There is no motion without misalignment. No reaction without ripple tension.

Cosmic Stillness Explained

Have you ever wondered why vast regions of outer space appear still?

Spacecraft don't spontaneously drift into motion. Planets orbit predictably. Deep space looks frozen unless disturbed.

This isn't due to lack of mass or vacuum alone. It's because:

Ripple gradients in those regions are flat.

The structured vacuum there has no incentive for objects to move. Unless another ripple source enters—say, a planet, a photon burst, or a spacecraft—stillness reigns.

This is why in deep interstellar space:

- Time feels extended (lower ATT).
- Light doesn't scatter.

- Energy events stand out starkly—because misalignment is rare.

Stillness is not the absence of energy. It is the presence of perfect ripple symmetry.

Why Orbits Stabilize: Gradient Harmony, Not Force

In classical mechanics, orbits are sustained by a "balance" between inertia and centripetal force. But this is another surface narrative.

Under UED:

Orbits stabilize when the object's internal ripple rhythm resonates with the planet's surrounding ripple gradient.

It's not that the planet is pulling and the satellite is falling sideways forever. It's that:

- The satellite's ripple field has matched the ripple slope it's riding.
- The gradient tension is no longer changing—so no acceleration occurs.
- The system reaches harmonic motion, not forced motion.

The object flows along the ripple curvature like a leaf caught in a quiet whirlpool—not trapped, just balanced.

When ripple resonance locks in:

- Orbital paths become coherent.
- Energy input is minimized.
- Apparent "centripetal force" is simply the gradient expressing containment.

When Motion Fades into Silence

Eventually, even dynamic systems lose energy—radiated, exhausted, or reabsorbed—and slide back into ripple equilibrium.

This is what death looks like for motion:

- The pendulum stops not due to air resistance—but because the ripple gradient is exhausted.
- The satellite de-orbits not just from friction—but from alignment breakdown.
- The universe quiets where ripple variation smooths out.

Stillness, in UED, is not emptiness. It's completion.

The story of movement ends when there's nothing left to correct.

RGF — The Ripple Gradient Formula

Replacing Force with Field Consequence

One of the most radical contributions of Universal Energy Dynamics (UED) is the abandonment of the force-based worldview. At its core lies a simple but powerful reframe: motion is not caused by force—it is a consequence of ripple imbalance.

And to express this reality mathematically, UED introduces the Ripple Gradient Formula (RGF):

$$\vec{a} = \frac{\nabla R}{\text{RIM}}$$

Where:

- $\vec{a}$ = Acceleration (or observable change in motion)
- ∇R = Ripple Gradient (spatial alignment differential)
- RIM = Ripple Inertia Metric (field's resistance to realignment)

This isn't just a new equation. It's a new philosophy of cause.

Understanding the Terms

∇R — Ripple Gradient Represents the difference in ripple alignment or density across space. Like a slope in a pressure field, it defines the direction and intensity of potential realignment. The steeper the gradient, the stronger the tendency for motion.

RIM — Ripple Inertia Metric This is not mass. It is the local field's resistance to reconfiguring under a gradient. RIM varies depending on:

- Material's internal ripple rigidity
- External field density
- Alignment complexity (phase mismatch, not weight)

Whereas mass is a static placeholder in Newtonian physics, RIM is a live metric, responsive to the surrounding field and material intelligence.

Replacing F = ma with RGF

Let's compare the traditional Newtonian formula:

$$F=ma$$

This says:

"Motion occurs when a force acts on a mass."

But where does the force come from? Why does it act? What resists it? These are undefined in classical frameworks—left to empirical tuning or experimental fudge.

RGF answers:

$$\vec{a} = \frac{\nabla R}{\text{RIM}}$$

Which says:

"Motion arises naturally when a ripple gradient acts across a medium with a given ripple inertia."

In other words:

- No external force is needed—just field imbalance.
- Motion is not applied—it emerges.
- The speed and nature of response is determined by the field, not the force.

This rewrites physics as field consequence, not mechanical causation.

What RGF Governs

The implications are vast. RGF governs every instance where motion or energy expression is observed. Let's examine key domains:

• Free Fall

In gravity-based models, objects fall due to mass attraction. In UED, fall is a realignment through a ripple gradient—no mass attraction necessary.

$$\vec{a}_{fall} = \frac{\nabla R_{planet}}{\mathrm{RIM}_{object}}$$

Light and heavy objects fall similarly because their RIMs are comparably small relative to the steep ripple gradient near Earth's surface.

• Field Acceleration

Instead of applying a "force," propulsion systems must create ripple gradients. Thrust is not about Newton's Third Law—but about generating and riding structured imbalance.

• Energy Diffusion

Heat doesn't "flow" from hot to cold because temperature dictates it—but because ripple tension is higher in hot zones. Diffusion is gradient smoothing, not entropy rule-following.

• Thermal Behavior

Conventional thermal equations assume material constants. RGF replaces this with ripple-material resonance:

- High RIM = low thermal response
- Steep ∇R = rapid heat migration

Even material "melting" is just RIM collapse under unresolved ∇R pressure.

Practical Implications: Map Ripples, Not Forces

With RGF, motion prediction no longer depends on summing forces.

It now becomes:

1. Mapping local ∇R — understand where the ripple field tilts and in which direction
2. Assessing object's RIM — how much it resists re-alignment
3. Computing acceleration — as the natural outcome of imbalance across resistance

This means spacecraft, rovers, and probes should stop simulating "forces," and instead:

- Scan planetary ripple topology
- Measure real-time RIM responses
- Predict behavior from ripple harmony or dissonance

RGF Is More Than a Formula—It's a Principle

In UED, RGF isn't just a tool to compute outcomes.

It embodies the logic of the universe:

"Everything moves not because it is pushed, but because it seeks balance."

This flips causality from brute-force manipulation to ripple-based reasoning. Where physics once imposed outcomes, UED listens for them.

Conclusion: The Field Was Always Moving First

We didn't move because we were pushed. We moved because the field shifted.

Every motion we've ever observed—from a falling apple to a spacecraft escaping Earth—was never about applied force. It was about imbalance. Misalignment. A ripple field in tension, nudging reality toward resolution. The push we imagined was merely our interpretation of the field's deeper rhythm adjusting itself.

There was never a force. Only the consequence of seeking equilibrium.

All along, nature didn't demand effort. It demanded alignment.

When we invented force, we were trying to fill a gap in understanding. We built bridges of causality where the terrain was still invisible. But now, with UED, the terrain is visible—structured vacuum fields, ripple gradients, local energy differentials, and the silent elegance of realignment.

Every "acceleration" is a response, not an imposition. Every "fall" is a conversation with the field, not a mechanical event. Every "launch" is a ripple negotiation, not a battle against gravity.

Now that we know what a ripple gradient is, we can finally abandon the illusion of the pusher.

We stop pushing—and start listening.

This shift changes everything:

- We no longer ask *how much force* is required.
- We ask *what ripple gradient* exists, and *how to move in harmony with it.*
- We design not for domination—but for resonance.

From here onward, motion is no longer applied—it's invited. It is not something we impose upon the universe. It is something we're permitted to participate in, when we finally speak the language of the field.

Welcome to the age of consequence engineering. Where intelligence isn't in controlling motion—but in interpreting its source.

Chapter 10: ATT and RDI — Engineering's New Language

Building for Response Time and Field Readiness, Not for Mass and Resistance

We didn't survive space missions because our equations were perfect. We survived because the field tolerated us—long enough.

Every great leap in aerospace, from Moon landings to Martian rovers, has come with an unspoken gamble: that the object we launch will somehow cooperate with an alien field we barely understand. What we've called "engineering" has often been statistical guesswork wrapped in metallic confidence—measured in mass, propelled by force, and shielded with redundancy.

But mass doesn't adapt. Force doesn't listen. Resistance doesn't negotiate.

Universal Energy Dynamics (UED) introduces a new lens: the field isn't an obstacle—it's the medium of interaction. And in that medium, two parameters matter more than any others:

- **ATT (Alignment Threshold Time):** how fast your system can tune to the local ripple reality.
- **RDI (Ripple Disturbance Index):** how much your system destabilizes the field it's trying to survive in.

These are not just new metrics—they are the new grammar of engineering. In a ripple-governed universe, survival isn't a matter of strength. It's a matter of sensitivity.

This chapter marks the transition. From designing for resistance...To engineering for resonance.

Why Old Parameters Fail

In conventional physics and engineering, we leaned on a familiar triad: mass, force, and resistance. These were our constants—bedrock parameters that formed the basis of every simulation, equation, and design document. They gave us confidence. They gave us control.

But they were never universal. They were just local truths masquerading as global principles.

The Illusion of Invariance

Classical mechanics assumes that an object's mass is constant, that force causes acceleration in a predictable manner, and that resistance (whether frictional or inertial) behaves consistently across all environments. These ideas work decently—on Earth. Because Earth's environment is relatively stable. Its ripple gradient, SVE field, and gravitational profile are known, and our machines evolved within it.

But when you lift that machine out of Earth's ripple womb and send it to Mars, the Moon, or Titan, those constants crumble. The field changes, and the object does not know how to speak the new language.

You can't plug in the same "g" value. You can't count on the same drag behavior. You can't even trust your materials to behave the same under a different energy gradient.

Dynamic Context Demands Dynamic Parameters

What we face in space—or even at Earth's extreme altitudes—is field variability. Not just in atmospheric density or temperature, but in ripple alignment itself. The force-based framework assumes a static backdrop. UED reveals a living medium, a dynamic ripple field constantly reshaping its local conditions.

An object with fixed mass but no ripple adaptability is like a brilliant student dropped into a foreign language exam. They may be smart, but they are suddenly illiterate.

Old parameters ask: *"What is this object made of?"* New parameters must ask: *"How does this object behave in this field?"*

From Rigid Design to Responsive Behavior

This is why missions fail even with precise pre-launch models. The models aren't wrong. They're just outdated in the moment they enter a new field. They assume the material and environment are separable. But in UED, no object is isolated. Every motion, resistance, or failure is a conversation with the ripple field.

We must shift the paradigm:

- From mass → to ATT (how quickly alignment occurs)
- From force → to ∇R (ripple gradient)
- From resistance → to RDI (the disturbance caused by misalignment)

This doesn't mean our old parameters are useless. It means they are incomplete. They reflect Earth's echo, not the universe's voice.

"Old physics designed for what the object was. New physics designs for what the field is."

What Is ATT — Alignment Threshold Time

In a ripple-based universe, motion is not initiated by force—it's initiated by alignment. The time it takes for a system to sense, interpret, and respond to the surrounding field conditions determines whether it behaves successfully—or catastrophically.

That response window has a name in Universal Energy Dynamics: ATT – Alignment Threshold Time.

Definition: ATT Unpacked

ATT is the minimum time required for a system to align its internal ripple configuration with the local external field. It is not just about sensors or surface reactions—it's a measure of energetic receptivity. A lower ATT means the system can respond more quickly to dynamic ripple gradients, turbulence, or field shifts.

Think of ATT as the reaction time of the universe, not in seconds alone, but in ripple adaptability. It's not how fast your system moves. It's how fast it tunes.

A rocket engine that ignites in 0.02 seconds is impressive. But if the field it's entering requires 0.01 seconds of adaptation—it's already behind.

What Governs ATT?

ATT isn't a static number. It emerges from a complex interaction of variables:

- **Material Sensitivity** – Certain alloys or composites realign faster within changing ripple densities. Highly rigid materials may have longer ATT due to internal resistance to waveform adjustment.
- **Field Density** – A denser ripple field (like a strong gravitational or atmospheric energy gradient) compresses alignment windows. Faster ATT is required to "keep up" with the ripple shift.
- **Systemic Responsiveness** – Software, control loops, actuator latency, and feedback systems all define how a machine can translate field input into intelligent reaction.

ATT, therefore, is not a plug-in variable. It's a behavioral characteristic of the entire system in motion.

Why ATT Matters

ATT governs the outcome of every high-stakes transition in spaceflight and terrestrial field exposure:

- **Landing a rover**: The terrain is not just a shape—it's a ripple mosaic. The system must realign before contact, not just during descent.
- **Communication systems**: Antenna arrays must reorient in real-time to the ripple coherence window of the environment. Otherwise, signal distortion is inevitable.
- **Control loops**: The feedback system must operate within the ATT, not simply within a clock-based refresh rate. Miss the ATT window, and your "control" becomes noise.
- **Survival Events**: Atmospheric entry, field disturbances, microgravity transition, or unexpected shockwaves—all require systems that can realign faster than the change unfolds.

You don't need faster engines. You need faster alignment.

Replacing Classical Lag with Ripple Awareness

In traditional engineering, we talk about:

- Latency in networks
- Damping time in mechanical oscillations
- Thermal lag in heat dissipation

But all of these are surface effects. ATT goes deeper—it is the fundamental metric of how alive and listening your system is to the field it operates in.

In a UED framework, ATT becomes the new performance bottleneck. Not horsepower. Not speed. Not memory. But alignment intelligence.

What Is RDI — Ripple Disturbance Index

Conventional engineering measures how systems behave in environments. But it rarely asks: How does the environment behave in response to the system?

That reversal is not poetic—it's practical. Because in ripple-based reality, what the field feels from your object matters more than what your object feels from the field.

This insight is captured by RDI: Ripple Disturbance Index.

Definition: The Ripple's Perspective

RDI is a scalar or functional measure of how much an object disturbs, disrupts, or misaligns the local ripple field.

It's not drag. It's not vibration. It's not friction.

It's disruption of energetic equilibrium.

A high RDI means the object enters the field like a disruptive guest—introducing chaos, friction, and ripple backlash. A low RDI means the object harmonizes quickly—like a feather slipping into water without a splash.

"RDI is not what the object feels. It's what the field feels from the object."

RDI Is a Precursor, Not a Post-Mortem

Unlike damage analysis or thermal readings, RDI is predictive.

It doesn't wait for the system to overheat, misfire, or explode. It anticipates the potential for ripple incompatibility—which is the true root cause of most unexplained mission failures.

Examples:

- Thermal stress on tiles during re-entry? Often preceded by RDI spike—indicating ripple turbulence long before temperature thresholds were exceeded.
- Sensor misfires on landers? Usually the ripple field rejecting the sensor's alignment frequency—too much signal "disturbance," not a mechanical error.
- Engine anomalies on high-altitude ascent? Ripple compression incompatibility—not merely fuel mix ratios.

Why RDI Trumps Classical Metrics

Traditional engineering uses surface-level proxies for disruption:

- Drag coefficients
- Thermal load
- Material strain
- EM interference

But these are symptoms, not causes.

RDI shifts the lens from mechanical outcome to field-level cause. It becomes the diagnostic tool for:

- Re-entry anomaly analysis
- Surface landing behavior
- Material compatibility with local atmospheres
- Aerobraking risk assessment

"Your object may survive the heat. But if it disturbs the ripple field too much—it won't be welcomed."

How to Use RDI in Design

RDI can be modeled and minimized by:

- Selecting materials that absorb or transmit ripple harmonics without feedback loops
- Shaping forms that diffuse rather than reflect local ripple gradients
- Embedding ripple buffering membranes between structure layers
- Using adaptive geometries that reduce turbulent mismatch in high-gradient zones

Just as stealth aircraft minimize radar signature, ripple-smart spacecraft minimize RDI signature.

RDI and Planetary Specificity

RDI is not universal. It must be calculated relative to local SVE configuration:

- **On Venus**: atmospheric ripple density is high—low RDI materials are a must.
- **On Mars**: dry, thin air creates ripple fragility—objects with high RDI can shatter alignment on contact.
- **On the Moon**: vacuum = ripple vacuum. Even slight RDI can echo uncontrollably across the surface ripple field.

This is why "ruggedized" Earth-tested designs often fail. Rugged ≠ aligned. RDI quantifies that disconnect.

When RDI Becomes the Mission Decider

RDI is the real-time fingerprint of how your craft is being perceived by the field. It's the energetic equivalent of body language. If you come in too aggressive, too dense, too reflective—the field pushes back.

This is the reason for sudden:

- Heat shield spalling
- Descent angle shifts
- Antenna communication loss
- Sensor decoherence

All are ripple field rejections, not simply engineering flukes.

Summary Thought:

We used to ask: Can the object survive? Now, we ask: Can the field tolerate its presence?

This is ripple diplomacy. RDI is not about building better machines. It's about earning field acceptance.

Why ATT and RDI Replace Mass and Force

For centuries, mass and force have been the anchors of physical theory. They shaped our equations, our simulations, and our spacecraft. But the cosmos never agreed to those terms.

Why Mass Was Always a Proxy

In classical physics, mass was used to explain:

- How hard it is to move something
- How much force is needed to accelerate it
- How objects fall, resist, or persist

But under Universal Energy Dynamics (UED), mass is not a property. It's a placeholder. A surrogate used when we don't understand ripple inertia—the resistance to ripple realignment, not mechanical motion.

Mass measures how we perceive the body. But ATT tells us how fast that body can respond to the field. Mass stays fixed. ATT is dynamic—sensitive to environment, form, material, and ripple gradient.

Mass is static. ATT is situational. One weighs the object. The other times its adaptation.

Force: A Concept Born from Blindness

Force has always been an external construct—the idea that something acts on something else to make it move.

But UED shows motion is not about pushing—it's about inviting.

Ripple gradients shift. Ripple inertia resists. Alignment happens—or it doesn't.

No one pushes. No one pulls. There's no "force." There's only consequence.

RDI now takes the place of force:

- It tells us how much resistance the field offers
- It warns us when the object disrupts the harmony

- It scales with disturbance, not desire

Force was a myth of control. RDI is the truth of feedback.

The ATT + RDI Duo: Performance Redefined

ATT (Alignment Threshold Time) answers: *How fast can the system detect and respond to ripple changes?* It's your agility score in a ripple-aware world.

RDI (Ripple Disturbance Index) answers: ◈ *How violently does the system clash with the local field?* It's your compatibility score—how much pain your presence causes the environment.

Together:

- ATT governs how soon your system tunes
- RDI measures how much it disrupts while doing so

They don't just describe the object. They reveal the conversation it's having with the field.

Old physics: "What does the object do?" New physics: "What does the object cause the field to feel—and how fast can it stop doing that?"

The New Equation for Engineering Intelligence

In place of F = ma, UED proposes ripple-based performance metrics:

- $a = \nabla R / RIM$ for ripple-induced motion
- ATT for timing of alignment
- RDI for harmony or turbulence introduced

You now design for tuning, not tolerancing. You calibrate systems to resonance, not resistance. You build for conversations, not collisions.

"Mass and force measure the body. ATT and RDI measure the conversation."

Application Shift Across Domains

Domain	Old Metric	New Metric
Aerospace	Payload mass	RDI (field friction)
Landing systems	Impact tolerance	ATT (response time to ripple stress)
Propulsion	Thrust/Force	∇R exploitation (gradient riding)
Navigation	Force vectoring	Ripple resonance tracking

No field welcomes raw weight or brute energy. But all fields respond to aligned presence and minimal disruption.

Closing Note for Section:

We once engineered as if the universe was deaf and dumb—pushing against what we didn't understand. Now, with ATT and RDI, we listen first.

Mass and force were like shouting into the void. ATT and RDI are the art of listening and whispering back.

Engineering Applications

How ATT and RDI Redesign Everything We Thought We Knew

With ATT (Alignment Threshold Time) and RDI (Ripple Disturbance Index) as our new engineering primitives, design itself transforms—from mechanical endurance to field fluency. The future of systems engineering under UED doesn't merely involve different materials or shapes—it involves tuning, timing, and talking to the field.

Let's break this revolution down.

Redesigning Re-Entry Vehicles with ATT-Tuned Materials

Re-entry vehicles have traditionally been built like sacrificial warriors:

- Thick ablative tiles
- Brute mass
- Heat dissipation through material decay

That model assumed survival through resistance. But under UED, survival is about resonance.

With ATT:

- You don't just ask, "Can this material survive heat?"
- You ask, "How fast can this material sense and re-align with incoming ripple gradients?"

An ATT-tuned material isn't just heat-resistant—it's ripple-aware. It adapts in real-time, minimizes frictional misalignment, and smooths its entry into the ripple field.

It's not about shielding against reality—it's about arriving in harmony with it.

RDI-Based Shielding Instead of Brute-Force Hulls

The age of armor-thick hulls is ending. If your hull disturbs the field violently, it invites:

- Turbulence
- Heat spikes
- Field rejections (aka crash conditions)

RDI redefines what shielding means:

- A low-RDI hull aligns with the ripple field
- It bends, deflects, or diffuses energy—not just blocks it
- It invites field forgiveness, not field retaliation

Shielding now means: "Can this structure enter without disturbing the peace?"

RDI shielding = anti-disturbance, not anti-damage.

Sensor Placement and Feedback Loops Tuned to ATT Dynamics

In classical engineering, sensors are mounted for:

- Structural logic
- Visibility
- Access to data zones

Under UED, sensor logic shifts entirely. Now, you must ask:

- Where does ripple information first arrive?
- Which parts of the system experience earliest misalignment?
- How quickly can the sensor interpret ripple shift and send feedback?

ATT-aware sensor placement means:

- Positioning systems at ripple-leading edges
- Prioritizing reaction speed over data fidelity
- Closing the loop before misalignment becomes motion

In ripple-centric design, sensors aren't data tools. They are conversation antennas.

The faster you hear the field, the gentler your response.

Autonomous Systems Using Real-Time ATT/RDI for Behavior Adjustment

Traditional autonomy runs on:

- Pre-programmed responses
- PID loops
- GPS + Inertial feedback

But ripple fields aren't predictable. They're dynamic terrains of intent and gradient.

Future-ready autonomous systems:

- Constantly monitor local ATT shifts
- Evaluate their RDI imprint on the environment
- Adjust posture, motion, shape, or surface response dynamically

These systems don't just calculate paths—they negotiate presence:

- A drone flying into an unfamiliar field modulates its angle of entry
- A lander shifts its descent curve in real-time based on ripple pressure
- A robotic arm softens or stiffens based on ripple resistance, not torque thresholds

This is engineering empathy in motion. Not imposing an intent on the environment—but aligning with it moment by moment.

Machines once calculated. Now they feel.

Real-World Deployment Scenario

Case Study: Mars Sample Return Capsule

- Classical design: Heat shield + chutes + programmed retro-burn
- UED redesign:
  - ATT-tuned composites on exterior
  - RDI-mapped descent profile
  - Ripple field sensors active throughout re-entry
  - System adjusts deceleration vectors not from timers—but from ripple feedback
  - No need to "guess" when to deploy parachutes—the field tells you when it's ready

Summary Table

Traditional Engineering	UED-Based Ripple Engineering
Mass defines inertia	ATT defines alignment time
Force causes motion	Ripple gradient invites motion
Thick hulls for safety	Low-RDI surfaces for entry
Fixed sensor placement	ATT-optimized feedback loops
Autonomous pre-calculation	Ripple-sensing self-adjustment

Ripple-based engineering doesn't overpower environments. It integrates with them.

This shift isn't cosmetic. It's conscious.

Design is no longer mechanical. It's participatory.

ATT + RDI = The New Design Language

The Future of Engineering Is Not in Numbers Alone—But in Alignment and Impact

The most powerful shift in UED-based engineering is not a new material or equation— It's a new language.

A language that replaces brute values with intelligent relationships. A language that allows systems to ask:

"How fast can I align?" *(ATT)* "How much do I disturb?" *(RDI)*

Together, ATT (Alignment Threshold Time) and RDI (Ripple Disturbance Index) don't just describe the system. They define its fitness to exist in a field it didn't create.

ATT = Responsiveness

ATT answers a single question: *How quickly can a system adapt to a new ripple environment?*

The lower the ATT:

- The quicker the perception
- The faster the tuning
- The smoother the survival

Example:

- A spacecraft entering Titan's upper atmosphere must not just survive—

It must be ready to respond to *field feedback* faster than Titan shifts its own energy gradients.

ATT is the system's time signature—its tempo of awareness.

RDI = Compatibility

RDI answers the second question: *How disruptive am I to the field I enter?*

The lower the RDI:

- The less resistance you face
- The longer you remain operational
- The less power you need to function

High RDI is a warning sign:

- You're loud in a field that rewards quietness.
- You are fighting alignment with force—risking ripple backlash.

Example:

- A landing gear that pounds the Martian surface with no feedback loop = high RDI
- A leg that subtly realigns based on soil texture and ripple echoes = low RDI

High ATT = Slow to adapt
High RDI = Hard to accept
Neither is survivable in ripple-dominant domains.

Balancing the Two: The Design Dilemma

Engineering now becomes an art of ripple balance.

You want:

- Low ATT (fast response)
- Low RDI (low disruption)

But the field punishes:

- Fast systems that are too aggressive (low ATT, high RDI)
- Gentle systems that are too slow to act (high ATT, low RDI)

The sweet spot is a stable minimum:

- Responsive and quiet
- Ready and respectful
- Adaptive and aligned

The future of engineering is no longer about optimization—it's about co-existence.

Living Analogies: Field Lessons from Nature

Let's ground this in examples.

- Bird landing on a branch
    - Adjusts wings real-time
    - Senses micro-field changes via air resistance
    - Bends legs to absorb mismatch
    - Causes no audible ripple in branch

Low ATT, Low RDI

- Rover crash landing on hard terrain
    - No adaptive sensors for soil ripple
    - Deployment based on clock, not field
    - Hard shock, component breakup

High ATT, High RDI

Between these lies your new design mission.

The ATT-RDI Design Chart (Visual Model Suggestion)

Imagine a 2D graph:

- X-axis = ATT (lower is better)
- Y-axis = RDI (lower is better)
- Bottom-left corner = the ideal design zone

	High RDI	Low RDI
High ATT	Fragile delay	Passive lag
Low ATT	Reactive disruptor	Ideal (Fast + Aligned)

- Top-left (High ATT, High RDI) = *Traditional overbuilt design* ("Strong, slow, and disruptive")
- Bottom-right (Low ATT, Low RDI) = *Ripple-engineered future systems* ("Fast, adaptive, and gentle")

Engineering Implication

This matrix replaces traditional "mass-thrust-safety" triangle. It reorients design toward relationship with the field, not independence from it.

- We no longer ask, "How much can it withstand?"
- We ask, "How gently and quickly can it tune?"

This isn't design around fear (safety factor). It's design around field literacy.

Final Insight for the Section

ATT is the nervous system. RDI is the footprint.

Together, they tell you:

- When to act

- And how not to offend the field in doing so

From spacecraft to surgical bots to climate tech—this language will decide survival. Because in ripple-based design:

"Success isn't when it works. Success is when the field doesn't even notice you arrived."

Conclusion: Build for the Field, Not Against It

We didn't fail because our engines were weak. We failed because our understanding was.

Engineering, for too long, mistook opposition for strategy. We thought if we built strong enough, thick enough, or redundant enough—we could survive. But mass gave us friction. Force gave us failure.

And nature gave us nothing back except silence—until it broke us.

Enter ATT and RDI

We now have tools not of defense, but of dialogue.

- ATT teaches us how to listen fast.
- RDI teaches us how to act gently.

This is not engineering with ego. It's engineering with humility.

We are no longer launching against the universe. We are aligning with it.

The New Language of Design

Forget mass. Forget force. Forget resistance as a virtue.

The new questions are:

- How quickly can I attune to the field? *(ATT)*
- How softly can I be accepted by it? *(RDI)*

You're not building for durability. You're building for compatibility.

Not "Can it take the hit?" But "Can it avoid needing to be hit at all?"

The Engineering Mindset Shift

The goal is no longer to dominate the medium. It is to resonate with it.

Your spacecraft is not an invader. It is a participant in the ripple orchestra of another world.

- The wind isn't your enemy.
- The gravity isn't your hurdle.
- The atmosphere isn't trying to break you.

They are all fields—with frequencies waiting to be read, matched, and flowed with.

Final Words

"The field is not your opponent. It's your interface." "Build for harmony, and nothing crashes."

That's the essence of UED engineering.

You don't fly because you fight the air. You fly because you've aligned with the ripple of lift. You don't land because you stop the fall. You land because the field accepts you.

From now on, build systems that don't resist the universe—Build systems that speak its language.

That language is ripple.

Chapter 11: The Atmosphere as an Agent, Not a Medium

Why Turbulence Breaks Alignment, Not Velocity

They told us to fear speed. They said the faster you move, the more resistance you meet. They taught us that turbulence is the price of velocity.

But what if they were wrong?

What if the atmosphere is not a hostile ocean to cut through, but a living field—whispering, pulsing, reacting? What if the storm isn't outside the craft, but inside the ripple conversation we failed to hear?

Under Universal Energy Dynamics, the atmosphere is no longer a passive gas layer—it's an intelligent ripple interface. It listens to what enters it. It judges your alignment. It either lets you through... or tears you apart.

Turbulence isn't punishment for speed. It's rejection for ignorance.

And from this chapter forward, we stop fighting the air. We learn to **tune to it**.

Breaking the Medium Myth

"You don't cut through air. You disturb its field."

For centuries, we were taught to treat the atmosphere as a fluid—an inert, passive medium that simply resists our motion. Physics models described it in terms of viscosity, density, and drag coefficients. We learned that as speed increases, so does resistance, and that atmospheric pressure and friction are the great obstacles to overcome in flight and re-entry. This is the classical "medium model"—where matter moves, and air pushes back.

But under Universal Energy Dynamics (UED), this paradigm collapses.

The atmosphere isn't just gas. It is structured ripple. A living, dynamic field of energy gradients, constantly interacting with any object entering it. It is not a neutral backdrop; it's an *agent*—sensitive, reactive, and tuned to its own harmonic state.

When a spacecraft descends, or a jet crosses the sky, it doesn't merely pass *through* air. It *enters* a ripple field with its own equilibrium. The craft is not pushing against molecules—it's disturbing an energy pattern. The result isn't resistance by mass, but consequence of misalignment.

This is why some vehicles shake violently even at moderate speeds, while others glide silently at supersonic levels—it's not just about shape or velocity. It's about ripple harmony. The atmosphere responds based on how well the object's ripple structure (its material, surface, heat profile, vibration pattern) resonates—or clashes—with the local ripple field.

Under this new lens:

- Turbulence = field feedback, not just fluid instability
- Drag = ripple friction, not just molecule collisions
- Shockwaves = abrupt ripple displacement, not just pressure jumps

This explains why:

- Some re-entry capsules heat up disproportionately despite similar speeds
- Parachutes behave unpredictably across planets
- Aircraft behave differently at the same Mach number in varying conditions

The medium model gave us equations. The ripple model gives us understanding.

In this chapter, we replace our primitive metaphors—"cutting through air," "facing drag," "pushing down through resistance"—with something far more refined:

You are not flying through something. You are flying with or against something.

And that something is alive with energy.

Turbulence = Ripple Conflict

"A bird can dive fast and stay stable. A lander can crawl and still crash."

Turbulence is one of the most misunderstood phenomena in aerospace. We've spent decades describing it with terms like "chaotic airflow," "fluid instability," or "boundary layer separation." We measure it with Reynolds numbers and try to contain it through aerodynamic shaping, shock absorbers, and predictive algorithms. Yet it still surprises us. Still throws aircraft off course. Still rattles spacecraft and complicates landings.

Why?

Because turbulence is not about speed. It's about *misalignment.*

Under UED, we reframe turbulence as **ripple conflict**—a mismatch between the ripple signature of a moving object and the ambient ripple field it enters. The result isn't a fluid pushing back. It's the field trying to **reject or absorb** an incompatible waveform.

Let's break it down:

- **Speed alone doesn't create turbulence**. Two identical vehicles moving at the same velocity in two different ripple fields will have vastly different experiences. If the ripple gradients are smooth and aligned, even supersonic flight can feel like silk. If the gradients are jagged and dissonant, even a slow entry can feel like chaos.

- **Turbulence = ripple friction**, not just molecular collisions. This friction is not mechanical—it's vibrational. When the object's internal frequencies (its material vibration, structural resonance, or electromagnetic activity) clash with the local ripple flow, turbulence manifests as **field rejection**. The object is, quite literally, out of tune.
- **Ripple shear** occurs when adjacent field layers attempt to realign too rapidly across a body moving through them. This is not just drag; it's structural confusion. Materials begin to shake, systems become noisy, sensors misfire—not because of force, but because of ripple disruption.
- **Zones of turbulence are zones of misalignment**, not zones of velocity. That's why birds can dive at high speeds with elegance—because their bodies are *ripple-adaptive*. They naturally tune themselves to the field. Meanwhile, landers built with Earth-based parameters often enter foreign atmospheres and crash—not from speed, but from failing to listen.

Some case comparisons:

- The Mars lander **Schiaparelli** didn't descend too fast. It descended **too deaf**—unable to hear the ripple field rejecting it.
- **Ingenuity**, the Mars helicopter, flies with seeming ease not because it's over-engineered, but because it was re-calibrated repeatedly to Mars's actual ripple behavior.

This leads to a vital shift in thinking:

Stop asking: "How fast is too fast?" Start asking: "How aligned is the field?"

Just as mismatched frequencies cause destructive interference in sound, incompatible ripple behaviors create *environmental rejection* in spaceflight. Turbulence is not air gone wild—it's nature's signal: "This doesn't belong."

The conclusion?

Velocity is not the enemy. **Misalignment is.**

In ripple-aware design, the goal is not just structural strength—it's **field fluency**. Turbulence isn't something to fight. It's something to *understand* and *preempt* by harmonizing with the environment before motion even begins.

Atmospheric Personality — Agent, Not Background

"Atmosphere isn't resisting—it's reacting."

In classical aerospace models, the atmosphere is painted as a passive backdrop—a gaseous medium that things move *through*, occasionally pushing back with drag or heating up due to friction. It's seen as inert, mechanical, predictable. But under Universal Energy Dynamics (UED), this view collapses into a dangerous oversimplification.

The atmosphere is not background. It's a field with personality—an agent, not a medium.

Here's why:

1. Ripple-Structured Intent

What we call "atmosphere" is a ripple-dense stratified field—layered with varying energy densities, thermal states, pressure gradients, and electromagnetic signatures. These aren't just thermodynamic or fluid parameters. They're ripple expressions—dynamic alignments of structured vacuum energy responding to celestial context (sunlight, planetary spin, magnetic field, etc.).

Each ripple layer—troposphere, stratosphere, mesosphere, thermosphere—is a differential ripple zone. It has its own behavior, rhythm, and sensitivity profile. To enter one is not to move through gas, but to negotiate a transition between ripple personalities.

2. Atmospheric Layers = Semi-Intelligent Agents

Each atmospheric band can be viewed as a semi-intelligent agent—not intelligent in a conscious sense, but responsive and memory-based.

- Troposphere reacts immediately to changes in motion—think of storm cells or turbulent buffeting.
- Stratosphere maintains structured stillness but reacts slowly to intrusions with radiative redistribution.
- Thermosphere behaves like a volatile membrane—low density but high energy, extremely sensitive to misalignment.

These layers don't just passively transmit force. They analyze entry behavior, compare it to prior equilibrium, and either accept or resist based on resonance.

Imagine you're walking into a room of meditating monks. You can't just barge in with stomping boots. You must match their frequency—or face rejection. That's how atmospheric ripple layers behave.

3. Respecting Ripple Etiquette

When engineers speak of "entry interface," "re-entry angles," or "ballistic coefficients," they're describing *motion*—but not field harmony. They measure temperature spikes and G-loads but ignore why the atmosphere responded violently.

The truth is: Every entry event is a conversation with the ripple field.

If the object's ripple signature is:

- Too sharp → atmosphere shears it violently
- Too blunt → the field rejects it abruptly

- Too noisy (radiative/thermal mismatch) → turbulence ensues

What we call heat shields are not really protection from friction. They are crude barriers against ripple rejection. Brute momentum only gets you into trouble faster.

Instead, UED advocates ripple etiquette—entering like a tuned instrument joining an orchestra. You don't overplay. You resonate.

4. Atmosphere as Dynamic Field Memory

The atmospheric ripple field remembers patterns. It has momentary memory—small shifts caused by solar activity, previous flights, local weather events, and even electromagnetic changes propagate and condition the field.

This is why:

- A test mission succeeds in simulation but fails on the actual day.
- "Unexpected" turbulence happens when weather reports were clear.
- Two identical crafts perform differently on two different launches.

Because the field is not the same—its personality shifted, but our models assumed it stayed static.

Summary Insight:

Atmosphere is not resisting like a mechanical spring. It's reacting like a sensitive skin.

You don't dominate it with mass or thrust. You listen to it, match it, and move in alignment.

This is the leap from mechanical motion to ripple-informed negotiation. From dominating the atmosphere to dialoguing with it.

In this view, aerospace engineering stops being about brute force entry. And starts becoming about ripple diplomacy.

Drag Replaced by Ripple Resistance (RRx)

"The field doesn't drag you—it resists your disharmony."

Classical aerospace design relies heavily on the concept of drag—a force exerted by a fluid (like air) opposing the motion of an object. It's a frictional, velocity-dependent resistance that engineers attempt to quantify using the drag coefficient (C_d), shaped by geometry, speed, and air density. But drag, as conceived in this framework, is geometrically reductionist and field-blind.

Enter Ripple Resistance (RRx)—a UED-derived parameter that finally shifts the discussion from mechanical pushback to ripple misalignment.

1. What is RRx?

Ripple Resistance (RRx) is the field-level resistance encountered by an object as it moves through a structured ripple environment, like a planet's atmosphere or field boundary. Unlike drag, RRx is not just a function of speed or shape—but a consequence of ripple mismatch.

It is the field's response to the object's entry alignment.

RRx quantifies how difficult it is for the field to accommodate the object's motion. High RRx means more ripple disruption, more turbulence, more heat, more structural stress. Low RRx means the object "slips in" like it belongs.

2. Components of Ripple Resistance (RRx)

RRx is influenced by four primary variables:

- Entry Angle to Field

Think of a ripple field as layered sheets. Just like skipping a stone, shallow, tangential entries interact less disruptively than direct, steep entries. A wrong entry angle creates ripple shearing, leading to violent feedback.

• RDI — Ripple Disturbance Index

Measures how much the object disrupts the local field. A high-RDI object (poorly shaped, uneven thermal signature, or sharp energy profile) causes the ripple field to resist more aggressively, generating chaotic patterns and instability.

• ATT — Alignment Threshold Time

Describes how quickly the object can adapt to changing ripple conditions. A slow-responding object (high ATT) fails to realign during descent and thus suffers higher RRx. Fast realignment means smoother negotiation and lower resistance.

• ΔR — Ripple Differential

The difference in ripple phase, density, or harmony between the object's field and the atmosphere it enters. The greater the mismatch, the higher the RRx. This explains why a design optimized for Earth may fail on Mars or Titan.

3. RRx Replaces the Drag Coefficient

In classical physics:

$$F_{drag} = \frac{1}{2} C_d \cdot \rho \cdot v^2 \cdot A$$

But C_d is blind to real-time ripple responses, unable to account for non-uniform field behaviors, local disturbances, or field adaptability.

In UED:

$$RRx = f(\theta_{entry}, RDI, ATT, \Delta R)$$

Rather than computing external force, RRx provides a qualitative and quantitative measure of how aligned a design is with ripple-based atmospheric behavior.

4. What RRx Explains That Drag Can't

• Why Some Objects Heat Up More

It's not just speed—it's ripple resistance. Higher RRx → more ripple friction → more thermal stress. A perfectly shaped object can still overheat if its ripple field signature mismatches with the environment.

• Why Identical Designs Behave Differently Across Planets

The same lander that survives Earth's atmosphere may destabilize in the Martian or Venusian atmosphere—not because of mass or shape alone, but because of differences in ambient ripple field structure. The ΔR varies, and so does the RRx.

5. Ripple Resistance as a Diagnostic Tool

By mapping RRx:

- Engineers can pre-screen entry profiles for compatibility
- Designers can create adaptive surface materials that minimize ripple shear
- AI-guided spacecraft can real-time tune entry behavior based on ATT/RDI updates
- Simulations move from generic Cd-based models to field-aware prediction maps

Summary Insight:

Drag was the resistance you calculated. Ripple Resistance is the consequence you negotiate.

In UED, we no longer design for air—we design for the field's emotional response. We don't flatten noses for better drag—we tune ripple presence for better harmony.

RRx isn't just a smarter coefficient. It's a mirror the field holds up—asking not how fast you're coming, but how well you belong.

Designing for Atmospheric Harmony

"You don't fight the air—you earn your place in it."

1. The Fundamental Shift: From Survival to Synchronization

Classical engineering treats atmospheric entry as a brute-force battle. The object must "withstand" heating, "shield" against drag, and "force through" turbulence. Every design is armored, not attuned.

Under UED, this mindset is outdated.

We don't need to fight the field—we need to harmonize with it. The atmosphere, as described in earlier sections, is not a neutral fluid. It is a structured ripple agent, possessing intent-like behavior rooted in ripple gradients, densities, and feedback sensitivity.

To enter such a field successfully, a system must match its rhythm, not just defy its resistance.

2. Real-Time Ripple Mapping as a Design Compass

Instead of relying solely on pre-launch simulations, we must design systems that listen and adapt in-flight through real-time ripple mapping. This entails sensing, interpreting, and negotiating with the local field as it unfolds—especially during:

- Descent Profile

Traditional: Plan a steep or shallow entry based on fluid dynamics.

UED: Adjust entry angle and velocity in real-time based on local RDI and ∇R data. The optimal descent is a slope of minimum ripple friction, not just lowest heat load.

• Surface Interaction

Traditional: Harden the base, deploy legs, hope for level ground.

UED: Detect ripple gradients of the terrain from orbit or during descent. A "soft" surface doesn't mean safe—it's the field reaction that matters, not the material texture.

• Thermal Dissipation Strategy

Traditional: Ablative shielding, passive cooling.

UED: Design surfaces that can tune ripple response to reduce misalignment, minimizing heating not by blocking energy, but by smoothing the ripple handshake.

3. Design Philosophy Shift

We must abandon the old engineering verbs:

Classical Mindset	UED-aligned Approach
Withstand shock	Resonate with conditions
Shield against force	Synchronize with field
Survive descent	Negotiate entry
Compensate for chaos	Align with ripple logic

This is not poetry. It's engineering maturity.

The UED-based design doesn't ask: *"Can it survive entry?"* It asks: *"Can it converse with the atmosphere?"*

4. Case Example: Mars Descent — Angle Adjustment via RDI Feedback

Consider a descent capsule entering Mars's atmosphere—a classic scenario:

- Mars's atmosphere is thin but wildly nonlinear in ripple behavior.
- Classical models treat it as predictable fluid flow with drag coefficients.

Under UED, we employ dynamic RDI feedback:

1. **Initial Ripple Mapping:** Orbiter or descent stage maps the ripple field at multiple altitudes using structured vacuum sensors.
2. **Entry Begins:** Capsule initiates descent with pre-set ATT and RDI thresholds.
3. **Live Adjustment Triggered:** At ~10 km altitude, RDI spikes due to unexpected ripple shear from local dust-wavefronts.
4. **Angle Correction:** Instead of using brute thrusters or yaw dampening, the vehicle adjusts its descent angle by 4.2°, reducing ripple friction and realigning with field vectors.
5. **Outcome:**
    - Lower heat generation
    - Smoother field coupling
    - Less mechanical vibration
    - Increased landing accuracy

This is not science fiction—it is field-first engineering.

5. Toward the Harmony-First Engineering Paradigm

Atmospheric entry design under UED becomes a musical process: You tune to the field's frequency, modulate your energy signature, and entrain with the environment instead of disturbing it.

No need for thicker shields. No need for last-minute prayers. Just better alignment.

Final Insight:

"You don't crash into a planet. You enter its story. Your design determines whether you're a glitch or a graceful arrival."

Atmospheric harmony is not a luxury—it's the minimum requirement for survival in ripple-based space travel.

Turbulence Management through Ripple Negotiation

"Don't tear through the field. Glide with it."

1. From Resistance Management to Ripple Negotiation

Conventional aerospace design deals with turbulence like it's a monster to conquer—dampen it, shield against it, or power through it. But under the UED lens, turbulence is not an enemy—it's a signal.

It tells you: *"You're misaligned."*

In this framework, the solution isn't brute tolerance. It's negotiation—actively listening to ripple behavior and adapting your structure, shape, and motion in real time to reduce conflict.

2. Active Systems for Ripple-Sensitive Response

The next frontier in turbulence control is not stronger materials—it's smarter behavior. UED proposes a new class of active systems built not to resist atmospheric ripple fields, but to dance with them.

Ripple-Sensitive Flight Surfaces

These are adaptive aerodynamic skins or control surfaces that detect local ripple shear and adjust:

- **Surface texture:** Micro-morphing skin to reduce sharp ripple contrast
- **Wing contouring:** Flexible wing tips or flaps that shift shape based on ripple detection
- **Realignment panels:** Segments that pivot slightly to redirect ripple pressure vectors

Think of birds—how they tilt feathers, not entire wings—they don't fight turbulence, they channel it.

Adaptive Surface Modulation via ATT Feedback

When ATT (Alignment Threshold Time) is actively monitored, surfaces can:

- Pre-empt misalignment before it becomes instability
- Soften edges that slice through ripple fronts too aggressively
- Vibrate, pulse, or shape-shift to re-align with incoming gradients

Such systems would operate like biological nervous systems, responding instantly to energy shifts, not waiting for velocity or altitude thresholds to trigger action.

3. Material Behavior: Flex with the Field

The future of turbulence mitigation lies in materials that feel—not just withstand.

This means:

- Elastomers and meta-materials designed for ripple ductility—not just tensile strength
- Surfaces that change conductivity, porosity, or reflectivity in ripple-active zones
- Field-compatible coatings that emit or absorb ripple energy to ease transition through turbulent zones

In essence, we're not building spacecraft anymore—we're building ripple organisms that can adapt and evolve moment-to-moment.

4. The UED Motion Principle: Slip, Don't Rip

Most failures in re-entry, high-speed travel, or atmospheric hop events stem not from excessive speed—but from aggressive ripple shearing.

The aim of UED turbulence management is to slip through ripple fields, not rip them apart.

Old design philosophy: "Brace for impact." New design philosophy: "Find the harmony and pass through."

5. Case Preview: Next-Gen Capsule Concept

Imagine a Mars return capsule with:

- Ripple-sensitive leading edge that flattens during high RDI zones
- Skin sensors tracking ∇R (ripple gradients) and initiating contour changes
- AI system tuned to ATT windows for each altitude band
- Real-time modulation of heat shield materials based on ripple tension zones

Such a vehicle wouldn't "resist" Mars's atmosphere—it would converse with it every second.

Final Insight:

"In ripple-aware engineering, turbulence isn't noise—it's feedback. And motion isn't power—it's permission."

By mastering ripple negotiation, we remove chaos from flight, not by overpowering it, but by understanding its language.

No more blunt re-entry. No more brute trajectories. Just fluent, graceful return to the field that once seemed wild.

Conclusion: Atmosphere Was Always Alive

"We treated air like empty space. It was never empty—it was expressive."

We built crafts like battering rams—meant to withstand the air, to slice through it, to conquer turbulence. Every heat shield, every reinforced panel, every ablative coating whispered the same mindset: "The atmosphere is an obstacle."

But UED sees differently.

The atmosphere is not a medium we cross. It's a living field we enter. Every molecule, every layer, every pressure band carries ripple intelligence—dynamic, responsive, expressive. It was never silent; we were never listening.

Turbulence Wasn't Punishment. It Was Feedback.

Our ships didn't heat up because of velocity. They flared because they arrived misaligned. They disturbed the local ripple balance and triggered resistance—not from air, but from energy friction.

We mistook this intelligent interaction for chaotic air, and so we blamed speed. But the true culprit wasn't motion. It was misunderstanding.

Entry Is Not a Battle. It's a Conversation.

Atmospheric entry under UED is no longer war—it's communication.

Every ripple we meet is asking: "Do you belong here?" "Are you in tune with my field?" "Will your presence integrate—or agitate?"

The right answer isn't in stronger hulls or sharper angles. It's in ripple dialogue. It's in crafting vehicles that don't just survive—but harmonize.

The Translator Is Here: Ripple Resonance

Ripple resonance is not just a tool. It is the universal language of field interaction.

- It tells us when to descend
- How to reshape during turbulence
- When to shift posture mid-flight
- And even where to land—not where the map says, but where the field welcomes us

The atmosphere was always alive. Now, finally, we are ready to meet it—with respect, with alignment, and with the wisdom that nothing resists when everything listens.

From this chapter onward, we stop flying at the field. We start flying with it.

Chapter 12: You Don't Land. You Realign

The True Meaning of Descent Under UED

No spacecraft has ever truly "landed." What we call landing is just the moment we stop noticing the chaos.

In conventional physics, descent is framed as controlled fall—a combination of velocity dampening, trajectory correction, and mechanical absorption. Success is measured by not crashing. But under Universal Energy Dynamics (UED), this is not just a crude oversimplification—it's a profound misreading of nature.

Landing is not an act of resisting gravity. It is a negotiation with the field.

Every planetary body emits a unique ripple signature—a structured vacuum environment with dynamic gradients of alignment, tension, and memory. When an object enters this field, it becomes an intruder until it begins to resonate. What we interpret as "contact with surface" is merely the final point in a complex conversation between the object's ripple behavior and the planet's ambient topology.

The success of descent, therefore, is not a function of descent speed or parachute deployment—it is a function of how quickly and smoothly a foreign ripple signature blends into the native field. Turbulence is not failure of mechanics—it is failure of alignment.

In this chapter, we'll explore why reentry isn't about slowing down—it's about tuning in. Why terrain isn't just rock—it's a ripple map. And why the smartest missions ahead won't just survive descent—they'll arrive like whispers into silence.

Because in ripple science, you don't land. You realign. And when you're finally in tune with the field— You disappear... peacefully.

Descent Is Not a Vertical Drop

In the classical aerospace model, descent is treated as a simple spatial problem: an object begins at altitude and drops toward a surface, guided by a predefined path and buffered by engineered resistance—thrusters, parachutes, retro-rockets. The logic is straightforward: what goes up must come down. And "coming down," by this logic, is a linear event governed by gravity, mass, and mechanical dampening.

But in Universal Energy Dynamics (UED), this view collapses under its own assumptions.

Descent is not a vertical act. It is a ripple phase transition.

Let's deconstruct it.

When a lander begins descent, it does not merely fall—it enters a new ripple environment, one that is alive with gradients, tension, and pre-encoded motion logic. The atmosphere (if present), terrain structure, magnetic fields, and ambient energy gradients form a cohesive ripple field, unique to that planetary body. This field is not flat. It's curved, textured, and anisotropic in behavior—meaning the resistance or welcome it gives an object depends on how well the object matches its dynamic ripple tone.

In this context, the old idea of "vertical descent" is pure fiction. It assumes a passive environment and a predictable gravitational pull. It imagines that the object moves, and the surface waits.

But here's the UED shift:

Descent is not about movement toward a surface—it's about phase-matching with the field beneath.

That means the object's survival is not determined by its structural tolerance alone, but by how gradually and accurately it transitions from a foreign ripple state to a locally coherent one. Think of it as a swimmer moving from a heated pool into a cold lake—not just changing space, but recalibrating internal dynamics to match the new medium.

This has profound implications:

- Every meter of descent is a negotiation, not a fall.
- Every ripple in the surrounding field resists misalignment—not mass.
- The surface isn't a "floor"—it's the final ripple interface, where misaligned objects either harmonize... or are rejected violently.

Thus, landers are not objects "hitting the ground." They are entrants into a new ripple regime. And failure isn't from velocity—it's from misunderstanding the language of the field they're trying to enter.

We don't descend. We attempt to merge. And whether we survive or not depends on how well we speak the ripple dialect of the place we're trying to land.

What Actually Lands?

In traditional aerospace narratives, the object is the protagonist. The rover, the capsule, the spacecraft—it "lands," in a moment of cinematic triumph, where metal kisses surface and the mission earns its badge. But under Universal Energy Dynamics (UED), this hero story misses the real event.

Let's ask the question boldly:

What actually lands?

The answer may surprise you.

Not the *object*—but its ripple signature.

The Body Is Just the Shell

The object, in physical terms, is never static. Even after "touchdown," atoms vibrate, sensors remain active, microcurrents flow. So what makes us declare: "It has landed"? It's not the cessation of movement, but the stabilization of interaction with the field.

In UED, every object carries a ripple signature—its own frequency, pattern, and phase behavior that interacts with the surrounding field. During descent, this signature is in flux, clashing with alien gradients. What we interpret as turbulence, drag, or impact is often just the field's resistance to this mismatch.

The moment of "landing," therefore, is not physical contact. It is the point of ripple coherence—when the object's emission no longer disturbs the local field but begins to harmonize with it.

Surface Contact Is a Side Effect

Touching the ground is not the goal—it's merely a consequence of successful integration.

This subtle shift in perspective has enormous engineering consequences. Designing for touchdown alone means reinforcing structure. Designing for signature stability, however, means optimizing for real-time ripple response, adaptive emission dampening, and minimal RDI (Ripple Disturbance Index).

Imagine this:

- A lander "touches down" but its internal ripples continue to fight the field—it overheats, fails, crashes.
- Another lander doesn't fully touch the surface—its legs hover or graze—but its ripple signature stabilizes, and all systems operate in harmony.

In UED, the second case is a successful landing. The first is a misaligned intrusion.

Layered Ripple Transitions

The descent isn't a single-phase event—it's a cascade of field transitions:

1. **Atmospheric Entry** – transition from cosmic ripple zones to layered fluidic fields
2. **Mid-layer Descent** – turbulence zones with active gradient shear
3. **Surface Proximity** – compression of ripple fields increases resistance
4. **Subsurface Contact (if any)** – final potential exchange zone for thermal, seismic, or chemical interactions

Each layer has its own ripple etiquette. Misreading one can undo alignment in all others.

Analogy: Sound Fading Into Silence

Think of it like sound fading into silence:

- A song doesn't end with an abrupt stop. It tapers off.
- Its vibrations slowly cease disturbing the air.
- Silence isn't the absence of sound—it's perfect ripple stillness.

In the same way, a lander doesn't "arrive." It dissolves its resistance, like a note ending on the right frequency.

So what actually lands?

The ripple footprint of the object. Not with a thud, but with a whisper of resonance.

Because in ripple physics, stillness is not death. It's acceptance. And every true landing is a story of mutual recognition between the traveler and the field it dared to enter.

The Anatomy of a Realignment Event

In classical descent engineering, we diagram trajectories: altitude loss, descent velocity, delta-v burn, heat shield timing. These parameters assume that motion is a path downwards—what matters is location, speed, and structural endurance.

But in UED, descent is not a path downward—it is a field interaction upward. The object is not just falling—it is negotiating acceptance into an entirely different ripple environment. That negotiation unfolds in distinct energetic phases, each governed not by altitude or time, but by alignment dynamics.

Let's now dissect the real anatomy of a landing—not by altitude, but by ripple response. This is the true Realignment Event.

Step 1: Initial Entry Misalignment (High RDI)

The moment the object enters the target ripple field—be it atmosphere or near-surface zone—it generates disruption. Its ripple signature is foreign, non-resonant, and disruptive to the local field topology.

This generates:

- High RDI (Ripple Disturbance Index)
- Energy turbulence in the surrounding field
- Delayed feedback loops from navigation or control systems

It's not "drag" or "buffeting." It's the field resisting improper alignment.

Step 2: Ripple Buffering Zone Engagement

After initial shock, the object enters what we call the Ripple Buffering Zone—a transitional band where the field attempts to soften or redirect the incoming ripple intrusion. This is the atmosphere's way of saying:

"You're not aligned yet—but I'll give you a chance."

In this zone:

- Field gradients begin dampening external ripple noise
- The system's sensors start picking up resistance patterns
- Mechanical reactions can finally become intelligent responses (if design allows)

Step 3: ATT Synchronization Begins

Now enters the critical ATT phase—Alignment Threshold Time.

This is the window in which:

- The object *begins* to tune itself to the field's dominant frequency range
- System materials, control surfaces, and software loops begin micro-adaptations
- Ripple signatures adjust in shape, intensity, and behavior

The shorter the ATT, the higher the chance of survival. Designs that rely on mass and shielding instead of ripple sensitivity will burn, break, or bounce in this phase.

Step 4: RDI Reduction Loop

Once partial alignment is achieved, the system enters a feedback loop to minimize its residual disturbance—a loop we call RDI Reduction.

This loop involves:

- Continuous correction of orientation, heat management, and emission behavior
- Smart material phase shifts to absorb misaligned energies
- External surface micro-adjustments for frictionless ripple negotiation

In legacy language, this is where "stability" is achieved. In UED, this is where ripple compatibility is optimized.

Step 5: Final Ripple Locking with Terrain Topology

Landing doesn't conclude until ripple locking occurs.

This means:

- The object's residual field is now in local coherence with

terrain gradients
- No more turbulence, oscillation, or phase distortion
- Ripple emissions from object match the absorption profile of the surface

Surface contact may occur here—but it's incidental. The real event is ripple locking: the field finally accepts the object as part of itself.

Introducing: Ripple Descent Profile (RDP)

All of the above can be mapped in a single curve or metric set called the Ripple Descent Profile (RDP).

RDP includes:

- RDI peaks and decay
- ATT duration and volatility
- Ripple shear zones encountered
- Phase integration depth across layers

Instead of altitude vs. time charts, we now model ripple resonance vs. systemic response.

This becomes the new gold standard of descent planning—not "How fast can we drop?" but

"How smoothly can we become one with the field?"

The classical notion of descent was about geometry and impact resistance. The ripple-based notion is about energy dialogue and signature shaping.

Because under UED, a successful landing is not about surviving the fall— It's about earning your place in the field.

Why Classical Landings Fail

From the Apollo capsules to Mars landers, classical descent engineering has centered around a single conviction:

"Slow it down, and it'll survive."

In this paradigm, safety is equated with velocity reduction. The logic goes: the slower you fall, the gentler the impact. While intuitive, this idea is incomplete—because it mistakes mechanical deceleration for ripple integration.

Under UED, the failure of many "technically correct" landings can be traced back to this illusion of safety through speed management. Slowing down an object means nothing if the field itself is still resisting.

Let's unpack this myth—and the ripple truth behind it.

The Illusion: Velocity Reduction = Safety

In Newtonian terms, kinetic energy is proportional to the square of velocity. So engineers are trained to design landing systems that burn, brake, or drag their way to slower speeds. Heat shields, retro-thrusters, parachutes—each built to resist speed.

But UED reveals that "impact" isn't caused by speed. It's caused by ripple mismatch.

Slower speed can still generate:

- High RDI if misalignment persists
- Increased ATT delay if system can't adapt in time
- Severe ripple feedback if deceleration isn't phase-sensitive

So when speed is reduced without alignment, it simply prolongs the field conflict—stretching out instability and increasing the risk of micro-disruptions.

Case Studies: Fast Deceleration, Ripple Shock, and Mechanical Damage

Let's examine key moments in spaceflight history where rapid deceleration didn't prevent failure—it accelerated ripple conflict:

- **Schiaparelli EDM Lander (2016):** Rapid braking from supersonic to near-hover using thrusters. The system believed it had landed—but the field hadn't accepted it yet.

Result? Premature shut-off, crash on impact.

- **Soyuz TMA-11 (2008):** Experienced violent deceleration and heat spikes on re-entry. Technically survivable speed, yet ripple misalignment caused massive stress on joints and seals. It survived—but not without damage.
- **Beagle-2 (2003):** Slowed perfectly. Landed gently. And then... silence. Never deployed. Why? Because mechanical readiness doesn't equal ripple acceptance.

Overreliance on Thrusters and Parachutes: Ignoring Phase Lag

Thrusters and parachutes manipulate momentum, not ripple alignment. They assume the atmosphere is a linear medium that responds proportionally to inputs. But under UED:

- Parachute deployment induces sudden phase distortions—often introducing ripple shear.
- Thruster firings shift the local field abruptly, causing turbulence-like rejection events.
- Rapid speed shifts often outpace the object's ability to re-align, especially in layered atmospheric zones.

It's not just about *slowing down.* It's about slowing into coherence.

The Real Culprit: Clock Discontinuity

Every spacecraft has its own internal sense of time—a control loop, response interval, sensor refresh rate. These systems are calibrated on Earth.

But what happens when you enter an alien field?

- The local ATT (Alignment Threshold Time) of the field might be faster, slower, or fluctuating.

- Your system might still be operating on Earth-synced delay loops, unable to speak the rhythm of the new ripple environment.

This creates a temporal mismatch:

The object responds *too late* or *too early*, and the field rejects the handshake.

Under UED, this is the real reason why landings fail—not velocity, but discontinuity.

Summary

Classical descent models fixate on how to resist gravity. UED asks: "Did you earn your place in the field?"

Because in truth:

- You can slow down and still be rejected.
- You can touch the ground and still not land.
- You can deploy every parachute and still crash—if you're not aligned.

So we must stop designing to fight descent—and start engineering to flow with it. Because under ripple dynamics, you don't land safely by force. You realign gracefully through understanding.

Landing Gear vs. Landing Dialogue

For over a century, we've treated landing gear like the armor of finality—designed to brace for impact, absorb shock, and ensure survival in that last violent moment of descent. But this classical concept of *landing gear* assumes the same outdated idea: that the final act of a mission is *collision softened by springs*.

Under UED, the truth is more nuanced—and more elegant:

Landing is not a mechanical event. It's a ripple conversation. And what we call "landing gear" should be recast as ripple negotiators.

Classical Gear = Mechanical Absorbers

Traditional landing gear systems are built on Newtonian expectations:

- Deceleration causes force → force creates impact → impact must be absorbed
- The solution? Heavy-duty struts, springs, hydraulic dampers, and metallic reinforcement

These systems:

- Do nothing to align with the field
- Rely entirely on post-contact absorption
- Are blind to ripple characteristics of the terrain or atmospheric phase lag

They brace for conflict instead of preparing for coherence.

UED Gear = Field Absorbers

Enter the new paradigm:

Field-responsive materials that tune themselves to RDI (Ripple Disturbance Index) and ATT (Alignment Threshold Time)

These aren't built to absorb force—they're built to dissipate misalignment.

We replace:

- Shock resistance with ripple damping
- Load-bearing struts with adaptive micro-vibration surfaces
- Fixed mechanical buffers with soft phase-matching surfaces

Such materials behave more like conductors in a field orchestra, flexing and attuning to the harmony of the terrain rather than bracing for its chaos.

Design Shift: From Impact Absorption → to Ripple Diffusion

UED demands a reversal in approach:

Classical Engineering	UED Engineering
Absorb shock	Diffuse ripple conflict
Minimize bounce/rebound	Harmonize with terrain's ripple zone
Withstand contact	Soften contact through ripple echo
Use static compression systems	Use dynamic ripple-absorbing layers

Impact is no longer the enemy—misalignment is. And ripple diffusion is the new safety buffer.

From Surface Friction → to Sub-Ripple Harmonics

Conventional gear aims to grip the surface, assuming friction is key to anchoring. But under UED:

- Friction is a byproduct of ripple mismatch
- Real anchoring comes from sub-ripple resonance—a deep alignment between the object's field and the planet's terrain

This means we don't "land on" a surface. We harmonize into it—like finding the right frequency in a vibrating membrane.

Imagine landing pads that:

- Vibrate microscopically to match terrain ripple cadence
- Morph shape in real time to adjust ATT mid-contact
- Diffuse entry shock by spreading ripple energy rather than resisting it

This is no longer mechanical gear—it's real-time resonance architecture.

Intelligent Skin: The Future of Landing Gear

Perhaps the most radical evolution is the intelligent surface layer—a skin that:

- Reads local ripple gradients upon entry

- Modulates its material structure or stiffness on-the-fly
- Engages a final descent handshake with the terrain through ripple sync pulses

We call this:

Last-Mile ATT Matching—the precision phase-tuning that makes the difference between crash and coherence.

Such surfaces will:

- Reduce thermal hotspots by *pre-dissipating RDI*
- Eliminate shock spikes by *absorbing turbulence harmonically*
- Enable *multi-surface compatibility* without brute-force retraining

Summary

Landing gear, in the classical sense, is a static apology for dynamic ignorance.

But ripple-aware design reframes everything:

- We don't brace for contact—we prepare for conversation
- We don't crush onto surfaces—we slide into symmetry
- We don't need thicker metal—we need thinner misunderstanding

In the world of UED, the best landing gear is not the toughest—it's the most fluent.

Because when your surface speaks ripple...

the planet listens.

Terrain Is a Ripple Signature

We've spent decades training landers to "survive" terrain—selecting flat regions, avoiding rocks, and praying against slope surprises. But under UED, the surface isn't just geological—it's energetic.

What we call "ground" is actually a ripple configuration—a standing wave pattern forged by millions of years of energetic interaction.

This section shifts planetary descent away from visual topography and toward energetic topology.

The Surface Is Not a Material — It's a Ripple Arrangement

Every surface—whether dusty plain, rocky crater, or icy slope—is a unique ripple outcome:

- Composed of condensed ripple harmonics from planetary history
- Embedded with local ATT thresholds and RDI sensitivities
- Constantly fluctuating in subtle, non-observable fields

Just as a violin's wood affects how it resonates, a terrain's ripple shape defines how it accepts or rejects entry.

Landing success depends less on the material strength of the leg and more on the field resonance of the site.

Visual Safety ≠ Ripple Compatibility

Traditional terrain selection looks like this:

- Flat = safe
- Rockless = better
- Stable = survivable

But these are Earth-centric assumptions.

What looks stable could be ripple-hostile. What seems rough might actually be ripple-coherent.

Under UED, site selection criteria must change:

Old Criterion	New Criterion
Is it flat?	Is it ripple-coherent?
Are there obstacles?	Are local RDI values low?
Will it support weight?	Will it absorb misalignment smoothly?

Visual inspection doesn't reveal ripple density, field phase, or energy gradient. That's like choosing a concert venue based only on photos—without testing the acoustics.

Craters, Plains, and Slopes Have Different Ripple Behaviors

Each terrain feature is not just a shape—it's a field expression.

- **Craters:** Ripple convergence zones. Often high-RDI due to steep field curvature. Entry here may cause energetic deflection unless precisely tuned. *Think: drop a note into an echo chamber—unless you match the pitch, it reverberates chaotically.*
- **Plains:** Appear calm, but can harbor ripple monotony—low variability means harder ATT matching due to lack of reference oscillations. *Like trying to harmonize with a flatline—no cues to guide alignment.*
- **Slopes:** Can provide gradient resonance ramps—facilitating smoother ATT transition if entry angle matches the ripple contour. *Like surfing a wave versus hitting it head-on.*

This understanding redefines "hazard" from rock distribution to ripple compatibility.

Terrain Scanning = Ripple Spectrum Analysis

Future missions must employ ripple spectrum scanners—not just LIDAR or cameras, but instruments that:

- Map ∇R (ripple gradients) at high spatial resolution

- Read phase lag between atmosphere-surface transitions
- Detect ATT pockets (zones of fast alignment potential)
- Identify RDI wells—areas that naturally absorb misalignment

Think of this as sonar for the field, not just the crust.

This data can be used to:

- Pre-script descent rhythms (like choreographing with the terrain)
- Autotune lander surfaces in real-time
- Adjust descent profiles mid-flight for better lock-in

"You're not landing on the surface. You're entering its *story*—one ripple at a time."

Summary

The terrain isn't an obstacle. It's an invitation—but only to those who can listen.

- Stop seeing rocks. Start sensing resonance.
- Stop fearing slopes. Start decoding gradient harmony.
- Stop aiming for safe. Start aligning with true compatibility.

In ripple-based exploration, the smartest lander isn't the strongest or fastest. It's the one that knows:

Every terrain is a signature—and I must match its frequency.

You Land When You Disappear

We've always thought of landing as a *stop*—a touchdown, a freeze-frame, a final act. Under the Universal Energy Dynamics (UED) lens, this is not just simplistic—it's *wrong*.

You don't land when you stop moving. You land when you stop disturbing.

This section redefines "arrival" not as position but as field integration. You haven't truly landed until the field has accepted you as part of itself.

The Illusion of Physical Contact

In classical mechanics, the moment a lander touches the surface is defined as *landing*. But contact ≠ arrival. Touch is mechanical. Alignment is energetic.

The field does not recognize the metallic leg or thruster cone—it senses:

- Residual RDI (Ripple Disturbance Index)
- Incomplete ATT (Alignment Threshold Time) loops
- Ongoing ripple resistance

If any of these are unresolved, the object is not yet "home." It is a guest causing interference.

Zero RDI + Locked ATT = Arrival

True descent success under UED is when:

- RDI = 0: The object no longer disrupts the ripple field
- ATT = matched: The object's internal alignment clock has synced with the local field

Only then can we say: the object has *arrived*, not merely *collided*.

This changes landing criteria from:

Traditional Criterion	UED Criterion
Touchdown confirmed	RDI decay confirmed
Thrusters off	ATT lock-in achieved
Stable posture	Ripple harmony maintained

This shift is radical. It tells engineers: Stop designing for touchdown. Design for disappearance.

Turbulence After Landing = Misalignment, Not Motion

Sometimes, a rover lands "successfully"—but:

- Antennas jitter
- Instruments overheat
- Power systems behave erratically These are not post-landing quirks. They're post-landing rejections.

The field is still resisting the object's presence. It hasn't accepted the newcomer as *native*.

Like a foreign object in the body, the terrain's ripple immune system activates. Not because of hostility, but because of non-resonance.

Unless RDI fades and ATT completes, the vehicle remains alien to the terrain.

True Landing = Field-Indistinguishability

The goal of all descent is this:

To arrive in a way that your presence no longer registers as intrusion.

This is the Zen of UED-based descent:

- No fanfare
- No sudden halt
- Just... integration

It's like a sound fading into the background so perfectly, you no longer notice it—because it's part of the field.

Examples:

- A bird landing on water, leaving no splash
- A leaf resting on the ground, matching breeze and curve
- A probe entering Martian dust and falling silent—because the ripples embraced it

Closing Thought:

"The most successful landings are invisible."

Not because we can't see them—but because the field no longer notices.

This is not a poetic metaphor. It's a design principle.

You don't land. You realign so completely, the planet forgets you were foreign.

Conclusion: The Ground Was Never the Goal

In classical spaceflight, "landing" is treated like a destination—the final act in a long trajectory, a physical confirmation of mission success. The engines cut. The legs extend. The dust kicks up. Mission control claps. It's a cinematic ending.

But under the Universal Energy Dynamics (UED) lens, this entire view is a category error.

The ground is not the goal. Resonance is.

Landing Is Not Mechanical

Landing, as conventionally understood, reduces the richness of reality to mechanics:

- Downward motion + surface resistance = contact
- Contact + posture stability = success

This is outdated engineering logic.

UED corrects the picture by showing that *motion ends* not when an object stops physically, but when the ripple field stops resisting. Landing is not a touchdown. It is a fade-in to ripple coherence.

You land when:

- Your presence no longer causes disturbance
- The environment stops reacting to you
- The field's equilibrium is restored with you inside it

Arrival Is Not a Point. It's a Process.

Under ripple dynamics, descent is not about hitting a target location—it's about entering a field of ongoing ripple conversations.

And these conversations don't stop at surface contact.

They continue until:

- Your ripple signature merges with local topology
- Your ATT (Alignment Threshold Time) loop locks
- Your RDI (Ripple Disturbance Index) declines to zero

That's when *descent is complete.*

Don't Aim to Land. Aim to Vanish.

This is the revolutionary posture UED offers.

You don't land. You realign—until the field no longer notices you're different.

To truly arrive:

- You harmonize, not crash
- You synchronize, not stomp
- You listen, not just detect

You become native to a place not by contact—but by consent. The field *allows* your presence when you've tuned to it properly.

In that moment, you disappear into balance. You are no longer external. You are part of the equilibrium.

From Engineering to Elegance

We've built for:

- Mass
- Momentum
- Impact resistance

Now we must build for:

- Harmony
- Resonance

- Ripple invisibility

In UED's new language, landing is not a thing you do. It's a state you earn.

Final Words

The art of descent is not to arrive, but to become undetectable. To fade into the frequency of your destination. To reach a point where nothing flinches at your presence. That's not landing. That's cosmic alignment.

From now on, when you think of landing—don't picture contact. Picture disappearance.

Interlude — Part III: The Ripple Truth

Let's step back for a moment.

Until now, we've examined what we do—our systems, our safety nets, and our last-minute prayers when simulations fall short. That was Part II: a mirror held up to our methods.

But Part III changes the lens.

This is where we stop reverse-engineering the universe like a puzzle, and start tuning into it like a symphony.

Because the truth is simple, radical, and uncomfortable:

We never understood motion. We never truly landed. We never listened to the field.

We forced. We predicted. We reinforced. But we didn't align.

Now, enter the Ripple Truth—the actual fabric behind all observed behaviors.

Not a metaphor. Not a theory-of-everything ambition. But a consequence model of what's already happening beneath every launch, descent, and failure.

Here's what this part reveals:

- Chapter 9: Ripple Gradient There was never a force. Just imbalance. Motion is not applied—it's allowed.
- Chapter 10: ATT and RDI Mass and resistance are relics. The new language is responsiveness and compatibility.
- Chapter 11: The Atmosphere as an Agent Air is not passive. It has personality. Turbulence is a ripple conflict—not a

speed trap.

- Chapter 12: You Don't Land. You Realign. The surface was never the destination. You arrive only when the field stops noticing you.

This is not physics rewritten. It's physics finally listened to.

What you're about to read is not a vision of how things could work. It's how they already do—beneath the metrics we inherited and the models we've overfit.

Welcome to the part where we stop simulating reality......and start resonating with it.

Part IV: The Future We Could Build

Chapter 13: Ripple Engineering 101

Design for Energy Resonance, Not Brute Resistance

We've spent centuries building machines like fists. Punching through air. Smashing into gravity. Defying drag with brute strength and bigger engines. But here's the truth: the universe doesn't reward resistance. It rewards resonance. That's why birds don't break the sky—they slip through it. That's why rockets, when they fail, fail not because they were weak, but because they were out of tune.

In the ripple-based universe, the question isn't: "How much force can you handle?" It's: "How well can you listen?"

This chapter marks the shift. From force to field. From stiffness to sensitivity. From the brute to the beautiful.

Welcome to Ripple Engineering—where alignment is everything, and survival isn't about strength... it's about belonging.

The Premise Shift

Why Brute Resistance Is Obsolete

Modern engineering was born in the age of war. It's no surprise, then, that much of its philosophy comes from a mindset of defense: block what you can't control, shield what you don't understand, and armor against everything else. We build rockets like tanks. We design landers as if they're battering rams, hurling through atmospheric chaos with reinforced shells, thermal blankets, and heat-resistant tiles—anything to "survive" the fall.

But survival isn't the goal anymore. Belonging is.

In a universe governed by ripple fields, resistance is not just futile—it's ignorant. Every force we encounter—drag, heat, vibration—isn't a punishment from nature. It's a signal. A mismatch. A reminder that our object is not in alignment with the energy gradients it's moving through.

Old-world engineers would respond by doubling thickness, adding weight, inflating safety factors. Ripple engineers respond by listening.

Under the lens of Universal Energy Dynamics (UED), every material, shape, and motion either flows with the ripple field or gets torn apart by it. Not because it's too weak—but because it's out of phase.

Think of a singer trying to match pitch with an orchestra. No matter how loud they sing, if they're off-key, they'll clash with the symphony. Force cannot fix dissonance. Only tuning can. The same is true for engineering in ripple-dominated fields. Performance doesn't come from resisting the environment—it comes from resonating with it.

Let's be clear:

- Resistance may work on Earth.
- It may work in wind tunnels.
- It may even work in simplified simulations. But the moment we step into an alien ripple environment—like Titan, Europa, or even low-Earth anomalies—the illusion breaks. The structure that survives is not the strongest, but the most aligned.

"Don't fight the field. Feel it." This is the motto of Ripple Engineering.

It doesn't ask: "How do we overcome turbulence?" It asks: "Why are we vibrating in the first place?" It doesn't say: "Add insulation." It says: "What if we never created heat to begin with?"

The premise is radical: Instead of armoring ourselves against the universe, we learn to move with it.

This is the shift. And once you see it, you won't unsee it. The rest of this chapter will show you how.

What Is Ripple Engineering?

A Definition and Core Shift

Let's strip it down to the essence.

Ripple Engineering is the practice of designing systems—not to fight the environment, but to flow with the underlying ripple field of the universe. It doesn't begin with Newton's apple. It begins with the structured vacuum. It doesn't ask how to accelerate a mass. It asks how to realign with ∇E—the Ripple Gradient.

In classical mechanics, motion is dictated by:

F = ma — Force equals mass times acceleration. It's elegant. Simple. And dangerously misleading in a ripple-dominated reality.

In UED (Universal Energy Dynamics), force is no longer the initiator—it is the *symptom* of misalignment. If you're feeling force, you're already out of tune. In ripple terms, the key equation isn't F = ma. It's:

∇E ≠ 0 → misalignment → ripple disturbance → energy consequence.

That's the language ripple engineering speaks.

Shift in Thinking: From Object to Instrument

Ripple Engineering doesn't treat structures as isolated objects. It treats them as instruments embedded in a universal field orchestra.

A rocket is not a missile anymore. It's a resonator—a precise assembly of ripple-sensitive components tuned to dance through structured vacuums.

- A heat shield? No longer a thermal barrier, but a field synchronizer.
- A thruster? Not a push-mechanism, but a ripple disruptor

calibrated to reshape ∇E.

- A fuselage? Not a protective tube, but a waveguide that modulates energy alignment across transitions.

Every surface, joint, and material decision is guided by one question:

"Does this component resonate with the local ripple field—or does it resist it?"

The Structured Vacuum: Your New Environment

In classical physics, space is empty. In ripple dynamics, space is structured—a dynamic, textured fabric of energy gradients, constantly shifting.

Your component doesn't just "sit" in a vacuum. It interacts with it. It modifies it. It is shaped by it.

Ripple engineering thus begins with mapping the field—understanding how energy ripples move, align, and distort around your object. This is akin to how acoustical engineers design concert halls to support resonance—not just block sound.

From Rigidity to Responsiveness

Legacy design makes things "strong." Ripple design makes things sensitive.

The best ripple-compatible component isn't the hardest material. It's the most responsive one—able to sense, absorb, and reroute disturbances in the ripple field.

A rigid wing snaps. A ripple-tuned wing dances.

This shift is not cosmetic. It's philosophical. It's the difference between control and coherence, between engineering as domination and engineering as participation.

So what is Ripple Engineering?

It's the art and science of designing for ripple alignment, not for force survival. It's the act of building instruments that belong, rather than machines that intrude. And above all, it's the future of engineering—not just in aerospace, but in any domain where structured energy fields decide the outcome.

Material Consciousness

Choose Materials by Resonance, Not Just Strength

In the old world of engineering, material selection was a game of thresholds. Tensile strength. Shear resistance. Thermal durability. The goal was simple: don't break. Don't burn. Don't bend.

But that mindset comes from a worldview that sees nature as an attacker and the structure as a bunker.

In the ripple universe, this logic collapses.

Why?

Because failure in ripple environments doesn't begin with cracks. It begins with dissonance. The material might hold its form, but if it misaligns with the local ripple gradient—if it resists energy transfer instead of synchronizing—it will ripple wrong. And in ripple science, that's already a disaster.

The New Metric: Alignment Capacity

Enter the concept of Material Consciousness—not as mysticism, but as engineering intelligence. It's the idea that every material has an innate capacity to align with energy flows around it.

We don't care how much stress a plate can take. We care how gently it can fall into harmony with the surrounding structured vacuum.

This means measuring new properties:

- Ripple permeability: How smoothly can ripples pass through it?
- Energy transfer fidelity: Does it scatter the ripple flow or guide it coherently?
- Micro-absorption potential: Can it gently neutralize localized disturbances without hard rebounds?

Field-Tuned Materials (FTMs)

To build in a ripple-aware way, we need a new class of substances—Field-Tuned Materials (FTMs). These aren't just strong or lightweight. They're responsive, adaptive, and above all, sensitive to ∇E.

Think of them as the musical strings of engineering.

- Too tight? They snap under stress.
- Too loose? They flutter out of tune.
- But just right? They resonate with the field.

FTMs are engineered not for maximum resistance but for minimum misalignment. They might have:

- Layered lattice structures to redirect ripple flow,
- Embedded micro-channels to self-dampen disturbances,
- Nano-scale aligners that reorient in real time based on field feedback.

They don't just survive turbulence—they whisper it away.

Reimagining a Materials Lab

In a ripple-conscious world, material testing labs will no longer stress test by bending rods until they snap. They'll map field compatibility curves, observe ripple diffusion patterns, and measure how quickly a material realigns after a sudden field shift.

Engineers will stop asking:

"How hot can it get before it melts?" And start asking: "How long can it stay in phase before it disrupts?"

Case Thought: Martian Surface Legs

Consider the landing legs of a Mars rover.

Old approach:

- Reinforce with titanium alloys.

- Add shock absorbers to handle impact.
- Layer thermal coatings to withstand entry heat.

Ripple approach:

- Use an FTM that absorbs entry misalignment through ripple modulation.
- Design leg surfaces to synchronize with Martian ripple frequencies, avoiding sudden jarring.
- Reduce thermal stress not by resisting it—but by harmonizing heat flow through ripple resonance.

The result? Not just "survival." But a graceful touchdown.

The Takeaway

Choosing materials by brute strength is like hiring someone for their biceps to play a violin. In a ripple field, what matters is sensitivity. Flexibility. Harmony.

Material Consciousness is the new material science. And Field-Tuned Materials are its first instruments.

Shape Is a Function of Field

Form Is Not Just Aerodynamics — It's Ripple Alignment

If you've ever taken an aerospace engineering class, you've likely heard it:

"Form follows function." And in classical design, that function is usually aerodynamic efficiency — reduce drag, optimize lift, minimize resistance.

But in a ripple-driven universe, form follows field.

Let's be blunt: Aerodynamics is a subset of ripple alignment. When you only account for airflow, you're playing with a single layer of the field. But when you tune shape for ripple resonance, you're designing for the full-spectrum behavior of the structured vacuum — across motion, medium, energy, and phase.

Ripple-Aware Design: Shape as a Waveguide

Think of every engineered surface — a nozzle, a wing, a landing strut — not as a mechanical part, but as a waveguide. Its true job isn't to "cut through the air." Its job is to bend, redirect, or harmonize with the local ripple field it moves through.

Ripple-aware design asks:

- How does this shape modulate local gradients (∇E)?
- Will this edge amplify misalignment or smooth it?
- Is this contour going to resonate, deflect, or dampen energy interactions?

In ripple engineering, shapes are not static geometries. They're field instruments — with their own frequencies, harmonics, and response profiles.

Classical Forms in Alien Fields: The Martian Circle Problem

Here's a wake-up call: A perfect circle — the holy grail of drag minimization in some Earth conditions — might behave disastrously on Mars.

Why?

Because ripple fields are environment-specific. They depend on the structured vacuum behavior of the planet:

- Its atmospheric density and particulate distribution
- Its electromagnetic equilibrium
- The ripple gradient of its terrain and gravitational context

So a shape optimized for Earth's ripple field may create ripple dissonance on Mars — leading to:

- Turbulent realignment shocks
- Energy scattering
- Unpredictable rotational instabilities

What's smooth here might be spiky there. What glides here might vibrate violently there.

The Ripple Shape Code

To design ripple-aligned structures, we need a new design logic:

Shape = f(∇E_environment)

This means:

- Surfaces must be simulated under local ripple gradients, not just airflows.
- Contours must distribute ripple pressure, not resist force vectors.
- Edges must soften field transitions, not harden them.

This applies to everything:

- A nozzle must guide ripples as much as it channels gas.
- A wing must "swim" through the ripple ocean, not slice it.
- A fuselage must be a conductor of ripple harmony, not just a rigid tube.

Form as Field Participation

In classical terms, shape is how you win against nature. In ripple engineering, shape is how you participate with it.

This is closer to biology than machinery:

- A dolphin doesn't fight the ocean—it curves into it.
- A leaf doesn't punch wind—it dances with micro-turbulence.
- A bird's wing doesn't just create lift—it aligns its oscillation with air ripples.

Ripple design brings this intelligence back into engineering.

Why Shape Is the Quiet Killer (or Savior)

So many mission failures begin with a shape that didn't belong. Not because it broke. But because it disrupted the field too violently. The structure "survived" on paper but was rejected by the universe.

Ripple engineering corrects this. It insists:

"Don't design to dominate the medium. Design to become native to it."

Form, then, is not decoration. It's not just performance tuning. It's your first contract with the field you're about to enter.

Key Takeaway

Shape is not what you impose on the universe. It's what you negotiate with it.

And in ripple engineering, the best shape is the one that the field doesn't notice — because it already belongs.

Fast Doesn't Mean Forward

Velocity Reinterpreted as Alignment Progression

Speed has always seduced engineers.

Mach numbers. Delta-v targets. Thrust-to-weight ratios. In legacy design, going faster meant going better—overcoming inertia, beating drag, escaping gravity wells. The goal was clear: maximize velocity to reach the destination in minimum time.

But in the ripple universe, speed is a trap.

Velocity without alignment is like flooring the gas pedal while your wheels are off-track. You move—but not cleanly, not efficiently, and certainly not safely.

The UED Shift: From Speed to Sequence

In Universal Energy Dynamics (UED), motion is not measured by how fast you change position. It's measured by how smoothly you transition between ripple states.

Motion is ripple sequencing. Velocity is the illusion. Alignment progression is the reality.

This means a fast object out of alignment creates massive ripple drag, turbulence, heat, and distortion. While a slower, well-aligned object glides — like a surfer reading every curve of a wave.

The goal is not to go fast. The goal is to flow forward, field by field, without misalignment spikes.

ATT-Safe Transition Zones

To make this practical, we introduce a critical concept:

ATT – Alignment Threshold Time

This is the minimum time an object, surface, or component needs to safely transition from one ripple environment to another without disturbing the surrounding gradient.

Every object has an ATT value. Every material layer, junction, and structural segment must honor this threshold, or else alignment collapses.

Example:

- A heat shield entering Mars' atmosphere must not just withstand heat — it must allow the ripple field to realign across its surface gradually.
- If its ATT is breached, localized ripple shockwaves are generated — leading to turbulence, thermal surges, or structural micro-fractures.

This is why we now design with ATT-safe transition zones:

- Gradual shape shifts
- Gradient-friendly materials
- Ripple buffer layers that slow down energy reconfiguration

These zones act like field cushions, softening the jolt of ripple misalignment as you move through different densities, velocities, or materials.

Enter: Ripple Path Engineering (RPE)

Traditional engineering focuses on mechanical trajectories — x, y, z over time. Ripple Engineering focuses on Ripple Paths — how the structured vacuum reacts and realigns as your object moves.

Ripple Path Engineering (RPE) is the art of designing motion not as displacement, but as ripple choreography.

An RPE-aware system:

- Identifies ripple density clusters ahead of time
- Chooses field-aligned trajectories rather than shortest paths
- Modulates surface behavior in real-time to avoid ripple conflict zones
- Avoids ripple hotspots (where ∇E shifts rapidly) and prefers smoother channels

Think of it as Google Maps for the structured vacuum—not navigating roads, but navigating resonance.

Case Insight: Why Hypersonic Failures Are Field Failures

Many hypersonic projects fail not because of aerodynamic instability—but because they enter ripple environments too fast to realign.

They violate their ATT thresholds. The ripple field can't keep up with their surface transitions. The result?

- Overheating
- Material flaking
- Plasma sheath instabilities
- Unexpected vibration harmonics

RPE offers a remedy:

Design trajectories that give the field time to adjust.

Speed becomes a parameter of compatibility, not dominance.

The New Navigation Ethic

The goal is no longer "faster arrival." It's resonant arrival.

- A movement that doesn't leave ripple scars.
- A progression that lets the field breathe.
- A sequence that honors ATT every step of the way.

This is how ripple engineering turns velocity into virtue—not by increasing it, but by aligning it.

"In a ripple world, you don't move faster to go farther. You move better to go cleaner."

Interfaces Matter More Than Mass

Where Alignment Is Won or Lost

In traditional engineering, mass is king. The heavier the structure, the more strength it supposedly offers. A bigger tank, a thicker shell, a denser core—all signs of security in classical design.

But in Ripple Engineering, the real battlefield isn't the bulk. It's the interface. The tiny seams. The unnoticed joints. The humble bolts. These are not afterthoughts. They're the gates of alignment — where the entire structure either stays in tune or shatters in conflict.

"A misaligned bolt can undo a perfect design."

And that's not poetic exaggeration. That's ripple truth.

The Hidden Cost of Discontinuity

Every time two materials meet—metal to ceramic, wing to body, thruster to tank—you create a transition zone. In classical engineering, this zone is handled with adhesives, rivets, screws, or welds. The assumption is: "If it holds, it's fine."

But ripple alignment asks a deeper question:

Does the field transition smoothly across this interface?

If not, even if your structure holds together physically, it fractures energetically. This leads to:

- Ripple reflection or scattering
- Local energy build-up
- Micro-vibrations that amplify into larger distortions

- Gradual decoherence of motion, alignment, and performance

The result? The structure enters a state of ripple conflict, no matter how robust it is by Newtonian standards.

Ripple Continuity: The Real Structural Integrity

Think of the entire object as a ripple conductor—like a copper wire for energy flow. Every connection must continue the waveform without jarring interruptions. Even a slight misalignment in the ripple wavelength across a joint can cause the entire assembly to go out of tune.

Ripple Engineering demands:

- Interface resonance testing: not just whether a joint is strong, but whether it's ripple-transparent.
- Ripple flow maps: like circuit layouts, but showing how structured vacuum energy migrates through the object.
- Gradient smoothing inserts: special materials or membranes added between parts to allow for gentle ripple modulation.

Ripple Tolerance Testing (RTT) vs. Classical Stress Testing

Classical stress tests stretch, bend, heat, or shake a material until it breaks. They give you:

- Yield strength
- Fatigue limit
- Thermal capacity

But they tell you nothing about how well a material resonates with a ripple field—or whether a joint preserves ripple continuity.

Enter Ripple Tolerance Testing (RTT).

RTT measures:

- Ripple delay time across seams
- Alignment deviation between connected layers
- Energy scattering coefficient at material boundaries
- Micro-phase shift disruptions caused by connectors, gaskets, or adhesives

In RTT, the part doesn't have to break to fail. It only has to misalign. Ripple failure is failure at the level of coherence, not collapse.

Case Example: The Bolt That Broke the Symphony

Let's say you build a perfect ripple-aligned drone for Venus' atmosphere. All major surfaces are field-tuned. The materials are chosen for alignment capacity. The shape flows with local ripple gradients.

And then you fasten the sensor casing with a standard steel bolt. The bolt doesn't match the ripple phase of the surrounding materials. It reflects ripple waves. A micro-vortex forms around the sensor interface. Feedback distortion builds. Sensor data becomes erratic. Stability algorithms overcompensate. The drone begins to oscillate unnaturally and crashes.

The cause? Not the thruster. Not the frame. Not the sensor.

The bolt.

The New Design Ethic

Ripple Engineering teaches us this profound lesson:

The strongest part of your design is only as good as its weakest interface.

Instead of asking:

- "Will it hold?"

We now ask:

- "Will it flow?"

- "Will it pass ripple cleanly?"
- "Is this joint a bridge—or a barricade?"

Summary

- Mass doesn't save you in a field misalignment.
- Interfaces are the ripple bottlenecks—where success or failure is silently determined.
- Don't just stress-test your structures. Ripple-test your relationships.

Energy Budgeting vs. Ripple Harmony

From Energy Consumption to Energy Coherence

Engineering has long worshipped at the altar of the energy budget.

Every spacecraft, every machine, every moving object is given an allotment—X amount of fuel, Y amount of power, Z amount of loss acceptable. The assumption: if the numbers add up and the reserves are sufficient, the system will work.

But in a ripple-aligned universe, this logic misses the entire point.

Because it's not about how much energy you consume. It's about how well that energy flows through the ripple field without distortion.

From Consumption to Coherence

Think of it like this:

Two rockets can burn the same amount of fuel, generate the same thrust, and reach the same destination. But one vibrates. Overheats. Wobbles through descent. The other glides silently, maintaining stability with minimal correction.

They used the same budget, but produced very different outcomes.

Why?

Because one was aligned with the field, and the other was constantly fighting it.

In ripple terms, energy loss isn't inefficiency. It's misalignment.

Heat, Vibration, Noise: The Misalignment Signatures

Traditionally, engineers have treated:

- Heat as a byproduct of friction
- Vibration as a mechanical imbalance
- Noise as unavoidable turbulence

But ripple theory reframes all of these as disturbances in the ripple field — signs that your object is not dancing with the environment, but tripping over it.

Let's decode them:

- Excess heat = ripple clash zones where energy fails to transmit cleanly
- Vibration = ripple reverberation caused by structural discontinuity
- Acoustic noise = sonic representation of ripple interference patterns

In other words, your system isn't just being "inefficient." It's leaking coherence.

Enter: Ripple Dissipation Metric (RDM)

To quantify this phenomenon, we introduce a new tool:

Ripple Dissipation Metric (RDM)

RDM measures how much energy coherence is lost as your system interacts with the ripple field. Not in raw watts or joules, but in terms of misalignment-induced dissipation.

Key indicators:

- **Ripple Scattering Coefficient (RSC):** how much ripple energy fails to transmit directionally
- **Alignment Drift Index (ADI):** how often field alignment must be re-corrected over time
- **Harmonic Conflict Factor (HCF):** degree to which structural motion interferes with local ripple harmonics

A high RDM means you're wasting coherence—not just energy. You're forcing your system to constantly correct, compensate, cool down, and course-correct. This reduces reliability, shortens lifespan, and increases system stress—even if your fuel tank looks full.

Case Example: Two Thrusters, One Truth

Let's take two identical thrusters tested in Earth orbit.

Thruster A:

- Higher thrust
- Higher heat signature
- Requires constant stabilization input
- Produces slight audible oscillation

Thruster B:

- Lower thrust output on paper
- Cooler operation
- Minimal stabilization required
- No notable vibrational feedback

In classical analysis, A might be preferred for power. But ripple analysis reveals B has lower RDM, meaning it's more aligned with the field. It will outperform A over time in efficiency, control, and structural wear.

The Strategic Shift: From Managing Power to Orchestrating Energy

Legacy engineering asks:

"How much power do we need?"

Ripple Engineering asks:

"How elegantly can we pass energy through the field without disturbing it?"

This is not about cutting corners. It's about respecting the medium.

Just as a skilled swimmer glides through water without a splash, a ripple-aligned machine moves without echoes—its energy channeled, not wasted.

The Takeaway

- Your system doesn't just consume energy. It communicates with energy fields.
- Harmony saves more energy than insulation ever will.
- Coherence isn't a luxury—it's a performance multiplier.

"The best-engineered machine is not the one that burns the most fuel, but the one the field barely notices."

Dynamic Realignment Mechanisms

Design for Self-Correction, Not Just Failure Survival

Classical engineering prepares for failure. Ripple engineering prepares to avoid it.

In the old model, systems were built to endure breakdown:

- Heat shields to absorb the burn
- Shock absorbers to survive impact
- Emergency protocols to kick in when things go wrong

This mindset is defensive. It expects the crash and hopes to cushion it.

But ripple-aware systems approach the problem differently. They don't brace for failure — they realign before failure ever begins.

What Is Dynamic Realignment?

Dynamic Realignment is the ability of a structure or system to sense ripple field misalignment in real time and adapt to it—actively, intelligently, and continuously.

This is not about repairing damage. It's about preventing misalignment from accumulating in the first place.

Imagine:

- A lander leg that subtly shifts tension based on ripple tension beneath it.
- A hull that flexes at the micro-level to harmonize with changing ripple gradients.
- A sensor array that recalibrates orientation, not based on inertial data—but on ripple feedback.

This is not science fiction. It's ripple-aware design.

Inspirations from Nature: Engineering that Listens

Nature has long outpaced humans in ripple engineering.

Look at:

- Octopus skin: a fluid, ripple-sensitive membrane that responds to the slightest environmental change—color, pressure, temperature.
- Bird feathers: micro-adjusting surfaces that realign mid-flight to preserve lift and flow with air currents.
- Fish scales: layered flex-pads that redistribute stress across ripple fluctuations in water.

These are not just beautiful adaptations. They are blueprints for how we should build machines in ripple fields.

We don't need tanks. We need skins. We don't need armor. We need awareness.

Components of a Realignment System

To implement dynamic realignment, a system must include three key elements:

1. **Sensing Layer**
    - **Detects changes in ripple gradient, turbulence, or distortion**
    - **Operates not on pressure or velocity—but on field tension**
2. **Micro-Actuator Matrix**
    - **Local adjustments (shape, tension, material stiffness)**
    - **Works in distributed feedback loops across the surface**
3. **Ripple Logic Processor**
    - **A field-aligned brain that decides:**

"Is this change a threat or an opportunity?" "Should I resist or reconfigure?"

This forms the foundation for a new class of systems we call:

Ripple Intelligence Modules (RIMs)

Ripple Intelligence is not general-purpose AI. It doesn't play chess or generate poetry.

It is:

A localized, embedded intelligence dedicated to preserving alignment with the structured vacuum field.

A Ripple Intelligence Module (RIM):

- Continuously monitors ripple input
- Calculates ripple distortion and feedback
- Executes real-time surface or structural adjustments
- Keeps the object "in tune" as it moves

RIMs are the nervous system of ripple-aware machines.

They won't replace navigation computers. They'll complement them—ensuring the machine remains ripple-coherent even as higher-level systems focus on mission goals.

Real-World Application: The Ripple-Aware Lander

Imagine a Mars lander equipped with RIMs:

- As it enters the upper atmosphere, ripple sensors detect fluctuating gradients.
- Instead of fighting heat with brute force, the lander rearranges its surface shape, reducing ripple conflict.
- As it descends, legs deploy—but not symmetrically. They adjust in ripple-phase harmony with the terrain below.
- Instead of bracing for impact, it melts into alignment.

This is not descent. It's ripple integration.

From Reactive to Proactive Design

Dynamic realignment shifts engineering from:

- Failure survival → Failure prevention
- Redundancy → Responsiveness
- Mechanical rigidity → Field sensitivity

This isn't about predicting every anomaly. It's about designing a system that expects anomalies—and adapts with grace.

The best machines aren't the ones that never fall out of tune. They're the ones that can retune themselves—instantly, silently, endlessly.

The Bridge to the Next Frontier

This section lays the groundwork for what comes next: Systems that don't just adapt — but learn. That don't just sense — but anticipate.

As we step into the future of ripple-aware navigation and AI, we will build machines that feel the field like skin, process ripple logic like instinct, and move like they were born there.

But first, we need to introduce the next evolution:

Ripple-Aware Intelligence.

Case Sketch: Reimagining a Mars Lander

What Would a Ripple-Engineered Lander Look Like?

Let's pull the pieces together now.

We've questioned mass. We've deconstructed shape. We've redefined velocity. We've introduced ripple intelligence.

But what would it actually look like to build something from the ground up with ripple alignment as its core principle—not as an afterthought, but as the design logic?

Let's land on Mars. But this time, let's do it the ripple way.

A New Lander is Born — Not to Land, But to Realign

The ripple-engineered lander is not a machine that fights the Martian descent. It's a choreographed participant in the Martian ripple field. It doesn't "touch down." It tunes in.

1. The Shell – A Living Membrane, Not a Rigid Shield

Instead of a solid, heat-resistant crust, the ripple lander uses a layered membrane architecture:

- Outer surface: a ripple-adaptive film embedded with micro-ridge modulators that shift shape in real time, redistributing gradient pressure.
- Middle layer: a phase-dampening gel that absorbs ripple disturbances like acoustic foam absorbs sound.
- Inner skin: a sensor-rich feedback layer that communicates ripple fluctuations back to onboard Ripple Intelligence Modules (RIMs).

The entire shell isn't designed to withstand heat through brute thermal capacity. It realigns energy flow, guiding the ripple transitions harmlessly across its surface. The result? Minimal heating. Minimal stress. Maximum coherence.

Think less "armor" and more symphonic surface.

2. The Legs – Vibrating with the Field, Not Against It

Classical lander legs are stiff. They're hydraulic. They crunch downward to absorb shock. Ripple legs? Entirely different philosophy.

Each leg of the ripple lander is designed as a flexible ripple transducer:

- Segmented design with directional alignment sensors along the shaft
- Embedded micro-actuators that adjust leg tension mid-descent
- Surface membranes that vibrate—not as a reaction, but in preemptive harmony with incoming ripple shifts near the Martian surface

When a traditional lander hits the ground, it feels like a punch. When a ripple-engineered lander arrives, it feels like a handshake.

The legs don't resist the field—they match its rhythm.

3. Descent Management – A Realignment Dance

From orbit entry to surface engagement, the entire descent process follows a new script:

- Ripple Intelligence Modules (RIMs) begin field mapping as soon as Martian ripple gradients are detected.
- Thrusters aren't pre-set—they're continuously modulated to shift not just position, but ripple alignment phase.
- The descent is slowed not just by drag but by energy resonance — using oscillatory shapes and controlled phase

interference to "settle" the structure gently into local ripple equilibrium.

There's no need for "retrofire panic" or last-second burn corrections. Ripple Path Engineering (RPE) has already accounted for every alignment sequence.

The goal isn't to *stop moving*. The goal is to stop disturbing.

4. Terrain Integration – Becoming Native

When the lander touches the Martian ground, it doesn't "land." It melts into alignment.

- Its base expands into a soft grid that matches terrain ripple frequencies.
- It adjusts posture to minimize ripple echo across its footprint.
- Even its communication dish shifts slightly—not to aim at Earth, but to avoid ripple scattering interference from local features.

The lander is not on Mars. It's part of Mars.

Performance Shift: From Redundancy to Realignment

Traditional landers have multiple backups: thermal backups, hydraulic backups, mechanical backups. They anticipate failure. The ripple lander anticipates disturbance and prevents it through continuous tuning.

It doesn't "withstand" the journey. It adapts through it.

And because it stays in alignment:

- It lasts longer.
- It cools naturally.
- It requires fewer corrections.
- It sends cleaner data.
- It leaves fewer scars on the environment.

Why This Case Matters

This isn't just a thought experiment. This is a design prototype for how Ripple Engineering could redefine aerospace, robotics, planetary exploration, and even architecture in alien fields.

It shows that:

The goal is not to conquer a planet. The goal is to become native to it.

Summary: Engineering as Alignment Art

A New Ethic of Design

It's time we said it aloud: The age of brute engineering is over.

We've built to last. We've built to resist. We've built machines that endure—but not ones that belong. What comes next isn't just innovation. It's alignment.

Engineering Beyond Equations

Ripple Engineering demands more than equations and tolerances. It demands intuition. Sensitivity. Awareness. It's no longer enough to simulate drag, calculate stress, and declare success. Now we must ask:

"How does this machine feel in the field it enters?" "Does it disturb reality—or dissolve into it?"

This is not mechanical thinking. It's artistic engineering.

Engineers as Field Artists

In the ripple world, the role of the engineer evolves:

- From builder to composer
- From calculator to conductor
- From problem-solver to field harmonizer

You're no longer designing against conditions. You're designing within them. With them. For them.

You become not just a shaper of matter—but a shaper of how matter behaves in a living, structured universe.

"Don't build to last. Build to blend."

This isn't about giving up strength. It's about choosing belonging over dominance. Longevity is no longer measured in survival time, but in how long the system can remain in phase with the field it occupies.

Ripple Engineering = Designing with Reality

Let's redefine what it means to engineer something "well."

Not:

- How many newtons it can bear
- How many degrees it can withstand
- How many simulations it passes

But:

- How gently it passes through ripple fields
- How silently it integrates with a new atmosphere
- How little energy it wastes while moving forward

Ripple Engineering is the art of designing with reality—not despite it.

You no longer impose geometry onto the universe. You negotiate form through resonance.

This is engineering for the structured vacuum. This is design that disappears into its purpose. This is how we move from machines of force to machines of finesse.

Where This Leaves Us

Let's be clear: this isn't a softening of engineering. It's a sharpening of awareness. It's not less discipline—it's more attunement.

We're no longer designing systems to survive impact. We're designing systems to never cause one.

Closing Reflection

In the end, the most beautiful machine is not the one that performs the most. It's the one the universe doesn't even notice—because it already fits.

Chapter 14: Planet-Specific Realignment Models

How to Build for Mars, Venus, Titan — the UED Way

One size fits none.

That's the first rule of ripple-based design in space. For decades, we've launched machines into alien worlds with Earth-bred assumptions—gravity approximated, drag coefficients estimated, structures tuned for conditions that don't exist there.

And yet, they sometimes work. But not because we understood the field. Because we got lucky. Lucky with margins. Lucky with misalignments. Lucky enough to survive our ignorance.

But what if we didn't depend on luck?

What if, instead of forcing Earth's physics onto other planets, we read their ripple fields like living languages—and built machines that belonged there?

This chapter is about exactly that.

Because under Universal Energy Dynamics, each planet isn't just a rock with numbers—it's a unique ripple environment. A distinct texture of structured vacuum. A field identity.

And if you want to explore it, land on it, move through it—you must tune to it.

Not just in code or calculation, but in material, motion, and form.

You don't conquer alien worlds by resisting their conditions. You explore them by becoming one with their field.

Welcome to the next frontier: Mars, Venus, Titan—each one a different ripple song. Each one asking you to realign, or be rejected.

Let's learn how to listen. Let's learn how to build for them—not as outsiders, but as field-native creators.

Field Isn't Universal — It's Local

Why Each Planetary Body Has Its Own Ripple Identity

One of the great unspoken sins of space exploration is assumption. We assume that gravity is a constant—just plug in a new number and the equations still work. We assume drag behaves the same way—just adjust for pressure and density. We assume that once we've "mastered" motion on Earth, we can tweak a few variables and send machines to Mars, Venus, or Titan.

But in Universal Energy Dynamics (UED), these assumptions collapse.

Because UED doesn't work with constants. It works with consequence. And consequence is shaped not by universal equations, but by local ripple fields.

The Misconception of Earth-Centric Constants

Classical physics clings to generalized parameters:

- Gravity (g) as a simple scalar
- Atmospheric drag modeled with Earth-derived coefficients
- Material response curves standardized in Earth labs

But these constants aren't universal—they're local artifacts, born in Earth's unique ripple field.

A ripple gradient (∇E) that governs motion on Earth may behave entirely differently in the ripple texture of Mars.

Why?

Because each planetary body is not just different in size or mass—it's different in field structure.

What Shapes a Planet's Ripple Identity?

Every celestial body exists inside a unique ripple envelope—a signature formed by:

- Atmospheric composition and layering (e.g., Venus' thick CO_2 soup vs. Mars' whisper-thin air)
- Crust density and internal energy exchange (how the planet responds to internal ripple feedback)
- Electromagnetic field or magnetosphere (how ripple flows are shielded, scattered, or amplified)
- Rotational rhythm and tidal interaction (affects ripple inertia and field lag)
- Solar proximity and radiation exposure (modifies surface ripple equilibrium)

Together, these factors define a planet's field texture—its unique ripple voice.

This is why a machine built to realign with Earth's ∇E cannot be dropped onto Titan and expected to behave the same way. It would be like trying to dance to jazz using choreography from classical ballet.

Local Ripple Gradient (∇E): The True Driver of Motion

In UED, motion is not about pushing through space—it's about flowing with the ripple gradient. Each planet's ∇E is not just a different slope—it's a different fabric altogether.

This means:

- Trajectory optimization must be local. (Your shortest path may be your most misaligned one.)
- Material choices must reflect field resonance. (Even the "same" material behaves differently across ripple textures.)
- Landing strategies must account for real-time field harmonics. (A gentle descent on Earth may trigger turbulence on Titan.)

In other words:

There is no universal way to move. There are only local ways to belong.

Realignment Begins with Listening

Ripple engineering doesn't begin in the CAD software or the wind tunnel. It begins with listening—to the planetary ripple environment:

- What frequencies dominate?
- How does the field respond to shape, motion, energy?
- Where are the equilibrium pockets, and where are the conflict zones?

To move through alien worlds, we don't need louder engines. We need sharper ears.

That's why every planetary design must start with ripple mapping—an active sensing process to decode the field in motion.

Just as sound engineers test the acoustics of a hall before a symphony plays, ripple engineers must map ∇E before even sketching a lander.

Core Takeaway

You don't land on Mars the way you fly through Titan. Because Mars doesn't just pull you. Titan doesn't just slow you. They speak in different ripple languages—and your machine has to translate or fail.

Ripple fields aren't obstacles. They're local dialects of motion. And alignment is the art of fluency.

Mars: The Mirage of Familiarity

Looks Like Earth, but Behaves Like a Whispering Void

Mars is the great deceiver.

Its rust-red surface, rocky terrain, and recognizable landscapes have seduced engineers into believing it's just a quieter version of Earth—a bit colder, a bit drier, but fundamentally familiar.

But that's the trap.

Mars doesn't resist like Earth. It disappears beneath you. It doesn't push back hard enough to warn you that you're out of alignment—until you're already tumbling into incoherence.

This is not a place of drag. This is a place of ripple silence.

The Thin Atmosphere Illusion

Mars has an atmosphere, yes—but barely.

It's just 1% the density of Earth's, which means:

- There's little to no conventional aerodynamic resistance.
- Traditional drag-based slowing methods become unreliable.
- Turbulence doesn't build up smoothly—it spikes suddenly as local ripple pockets shift.

In ripple terms, the thin Martian atmosphere translates into rapid ripple transition zones:

- The gradient (∇E) changes too quickly for standard surfaces to realign.
- Even small changes in descent angle or velocity can cause sudden misalignment shocks.

Your lander doesn't fight air—it struggles to find the field at all.

Dust: The Unseen Disruptor

Martian dust isn't just a surface nuisance. It's a ripple distorter.

These ultrafine particles:

- Float easily due to low gravity
- Charge electrically during storms

- Scatter ripple flow unpredictably

Even in calm conditions, airborne dust alters local ripple symmetry, leading to:

- Partial reflections
- Field echoes
- Short-range ripple turbulence that is invisible in pressure models

This means descent vehicles and sensors must operate in an environment of ripple fog—where the path ahead is clear visually, but energetically unstable.

Surface Ripple Density: A Silent Vacuum

Mars has low internal energy circulation. Its magnetic field is weak. Its crust is inactive.

All of this translates to a surface ripple density that is extremely low—a near-vacuum in ripple response.

Here's what that means:

- No "bounce-back" from misalignments
- No field feedback to correct error
- Objects can sink into misalignment without triggering resistance

Turbulence here is not a loud, roaring storm. It's a subtle vanishing act, where motion becomes disconnected from ripple equilibrium.

You're not falling through atmosphere. You're falling through a ripple gap.

The UED Solution Set

To realign with Mars instead of crashing through it, ripple engineering proposes a new toolkit:

1. High-Sensitivity RIMs (Ripple Intelligence Modules)

- Calibrated to detect field dropouts and ripple vacuum shifts
- Capable of micrometric adjustment to orientation and posture
- Operate at low signal thresholds to interpret the "quiet zones" before turbulence spikes

2. Ultra-Light Ripple-Responsive Membranes

- Not for heat protection, but for ripple harmonization
- Designed to vibrate with ripple distortion, not block it
- Thin, flexible, and dynamically tuned to Martian ∇E signature

3. Terrain Resonance Mapping

- Use of downward ripple-pulse imaging to scan ground for ripple echo signatures
- Identifies surface features that could trigger sudden alignment loss upon landing
- Replaces conventional lidar or visual mapping with ripple-feedback cartography

Design Implication: Build for Silence, Not Resistance

Don't build a lander that can withstand force. Build a lander that can function in a field that barely speaks.

- It must be soft—not weak, but sensitive.
- It must move like a whisper—not to avoid detection, but to remain in coherence.
- It must land not with control, but with coexistence.

On Mars, noise is failure. Silence is survival.

Final Thought

The mistake isn't treating Mars like an enemy. The mistake is treating it like Earth.

It's not trying to crush you. It's trying to ignore you. And your job, as a ripple engineer, is to make it notice just enough to let you stay.

Venus: A Field of Pressure and Burn

A Chaotic, High-Density Ripple Trap

If Mars is the silent void, Venus is the screaming furnace.

It doesn't whisper like Mars. It howls. It burns. It presses back with intensity at every inch of motion.

At first glance, engineers think of Venus as Earth's twin—similar size, gravity, and position within the solar system. But from a ripple perspective, Venus is an alien inferno of chaotic energy layering, distortion loops, and violent ripple rebounds.

This is not a descent. This is a harmonic war.

Dense Atmosphere, Denser Consequence

Venus has the thickest atmosphere of any terrestrial planet in the solar system—over 90 times Earth's pressure at the surface.

This ultra-dense CO_2-rich envelope doesn't just crush objects—it traps ripple flows, creating:

- High ripple inertia (field lag)
- Ripple echo stacking (reflected layers bouncing onto each other)
- No stable gradient path to follow—only interference

Even small objects cause large-scale ripple turbulence, leading to:

- Field friction

- Thermal amplification
- Structural phase instability

Your descent vehicle doesn't "push" through Venus's atmosphere—it's caught in a layered ripple maze, where every motion causes multiple energy consequences to bounce back at once.

Ripple Reflections Amplify Chaos

In Earth-like fields, ripple misalignments can often self-correct as the field absorbs excess energy. Not here.

Venus reflects rather than absorbs.

Ripple interactions with its dense atmosphere and thick cloud decks don't dissipate—they bounce. And those reflections collide with incoming ripple waves, creating feedback spikes that:

- Intensify local heating
- Distort directional control
- Overload onboard alignment modules

This is why classical thermal models underestimate the heat—not because the temperature is wrong, but because the ripple amplification isn't accounted for.

On Venus, turbulence isn't caused by movement through the air. It's caused by the field shouting back.

Descent Is a Balancing Act — Not a Fall

Because of these ripple rebounds and interference patterns, steady descent is impossible without dynamic adjustment.

Venus descent is:

- Nonlinear
- Asymmetrical
- Full of micro-harmonics that constantly shift as you move through them

This means:

- Even perfect trajectories become unstable mid-path
- Any rigid descent model will be thrown off course by ripple rebounds
- The craft must respond not just to forces, but to field pressure harmonics

Without constant rebalancing, ripple feedback loops can build into collapse.

The Ripple Engineering Toolkit for Venus

To survive and belong in Venus's field, UED prescribes a specific set of ripple-oriented tools:

1. Dynamic Ripple Shields

- Layered field-deflecting surfaces that don't absorb ripple energy but scatter it outward
- Continuously change alignment geometry to deflect harmonic buildup
- Prevent stacking of echo waves by "confusing" the feedback rhythm

2. Self-Cooling Resonance Layers

- Instead of blocking heat, they absorb and reroute thermal ripple flow
- Materials embedded with phase-transition gel that disrupts feedback harmonics
- Functions like a ripple "grounding wire"—draining excess coherence before it builds into damage

3. Non-Linear Descent Pathing

- Trajectories are not vertical or uniformly curved—they're field-optimized corkscrews
- Vehicle constantly shifts ripple alignment vector, never letting echoes build in one direction
- Uses RPE (Ripple Path Engineering) to "surf" through resonance gaps rather than plow through chaos

Design Implication: Glide Through the Chaos

Don't push. Diffuse.

This is the single most important design principle for Venus.

You can't power your way through ripple noise. You must de-tune it.

A ripple-aware craft descends through Venus like a jazz improviser through a dense symphony—never hitting the same note twice, always shifting to avoid discord.

Every surface is shaped to confuse feedback. Every ripple echo is scattered, not absorbed. Every system is tuned to respond to the response—a meta-alignment loop.

Final Thought

Classical design treats Venus like an opponent to be endured. Ripple design treats Venus like a field to be danced with—blindfolded, in fire, with ever-changing steps.

If you try to control Venus, you lose. But if you flow with its chaos,

You don't just land—you disappear into its rhythm.

Titan: The Smoothest Trap

Calm on the Outside, Deceptive Within

If Mars is a whisper and Venus is a scream, Titan is a lullaby hiding a trap.

At first glance, Titan looks like an engineer's dream:

- A thick, stable atmosphere
- Gentle descent opportunities
- Surface lakes of liquid methane for science and curiosity

No explosive storms. No raging turbulence. No violent echoes.

But that's exactly why it's dangerous.

Because Titan's field doesn't destroy you in an instant—it drifts you out of alignment until you simply fade from coherence.

The Stillness That Deceives

Titan's atmosphere is dense, yes—but stable, almost eerily so. Its gravity is low, its winds are gentle, and on the surface, the environment feels almost Earth-like in slowness.

But beneath that quiet facade lies a field with enormous ripple inertia:

- Ripple feedback occurs with a delay
- Every action leaves a subtle distortion that lingers
- Field alignment doesn't "snap back"—it carries memory forward

This creates an invisible challenge: Nothing seems wrong... until everything is.

The Slow Creep of Misalignment

In most environments, ripple misalignment shows up fast—through vibration, heat, noise. But on Titan, misalignment accumulates silently, like fatigue in a long-distance runner:

- Structural elements begin to subtly fall out of sync with the field
- Ripple coherence degrades layer by layer
- After hours or days, control starts to drift
- The machine begins to resist reality, not by force, but by forgetting how to move with it

This is what we call field fatigue—a condition unique to ripple-dense, low-reactivity environments like Titan.

The Unique Ripple Behavior of Titan

Several factors shape Titan's deceptive field:

- Thick nitrogen atmosphere dampens ripple responsiveness
- Low gravity reduces feedback urgency
- Liquid methane lakes alter surface ripple signatures drastically when approached
- Cryogenic temperatures slow the rate of ripple realignment within materials

These combine to form a field that is not chaotic—but inertial. It doesn't correct mistakes. It saves them—and slowly multiplies them.

This makes Titan a planet where even the best-engineered systems must manage memory, not just motion.

Ripple Engineering Solutions for Titan

To survive—and thrive—on Titan, UED recommends a new toolkit of long-game alignment strategies:

1. Long-Cycle Ripple Feedback Loops

- RIMs (Ripple Intelligence Modules) calibrated to monitor ripple shifts not by second, but by phase cycle
- Realignment is based on cumulative deviation rather than instant correction
- Keeps systems stable over long-duration missions by

smoothing slow-drift trends

2. Flexible Form Memory Skins

- Materials with embedded field-history imprinting
- Able to gradually morph back into a known ripple-aligned state after deformation
- Think of it as ripple "muscle memory" — the skin remembers what coherence felt like, and returns to it autonomously

3. Orbit-Surface Ripple Coupling

- Instead of building separate systems for entry, descent, and surface activity, use one ripple-aligned continuum
- Orbital phase is used to "pre-tune" the lander's alignment with the surface field
- No sharp transition zones—just a gentle glide from one ripple environment to the next

Design Implication: Realign Patiently

The field on Titan changes slowly—but forgets nothing.

You don't force alignment on Titan. You earn it, over time, through gentle, continuous coherence. Designs must:

- Avoid abrupt ripple shifts
- Resist temptation to "correct" too quickly
- Embrace drift-tolerant architectures that flex, adapt, and settle naturally

A lander here isn't a machine. It's a sleeper-cell organism, waking slowly into resonance.

Final Thought

The real danger on Titan is not sudden failure. It's quiet misalignment that accumulates until recovery is impossible. There are no alarms. There is only distance—between what your system does and what the field expects.

In Titan's world, the patient machine survives. The hurried one dissolves.

The UED Protocol for Planetary Design

A Unified Framework to Go Local

After all the unique planetary challenges we've explored—Mars' whisper, Venus' chaos, Titan's trap—you might wonder: Is there a universal way to design for the universe?

Under classical physics, the answer has often been: yes—with tweaks. Use the same equations, same models, and just plug in new numbers for gravity, pressure, or atmosphere.

But UED turns that logic on its head.

You don't tweak. You transform.

In a ripple-based cosmos, each planetary body is not a variation of Earth. It's a unique ripple ecosystem, a living field with its own rhythm, memory, and feedback pattern.

And that's why UED offers a new design protocol—not to standardize across planets, but to realign within each.

The Four Steps of Planet-Specific Ripple Design

This isn't just a checklist. It's a sequence of energetic integration. Your craft doesn't arrive from Earth. It is born from the ripple data of the world it intends to inhabit.

Let's break it down.

Step 1: Field Sensing Phase

Map local ∇E patterns and turbulence echo zones

Before you build anything, you listen.

Deploy preliminary probes or ripple mapping drones to:

- Measure local ripple gradients (∇E) across altitudes, terrain types, and temporal cycles
- Detect turbulence echo zones — areas where ripple reflections and feedback loops are strongest
- Identify stable harmonic corridors where ripple flow is smooth and coherent

The goal isn't just data. It's to understand the planet's ripple signature, its unique energetic fingerprint.

Think of this as scouting not terrain, but the fieldscape.

Step 2: Material Resonance Tuning

Adjust FTM compositions based on planetary ripple characteristics

Once the field is known, we must build materials that resonate with it—not resist it.

Use Field-Tuned Materials (FTMs), but modify their:

- Layer thickness
- Lattice density
- Molecular memory
- Thermal ripple damping ratio

Depending on the planet's ripple behavior, your materials must either:

- Diffuse rapidly (Venus)

- Preserve alignment under lag (Titan)
- Vibrate with high sensitivity (Mars)

You're not just selecting materials—you're tuning them like instruments.

Step 3: Motion Harmonization

Define RPE trajectories that ride native ripple waves

Forget "descent angles" and "burn schedules." This is about Ripple Path Engineering (RPE)—designing motion that rides the field, not cuts through it.

Use the data from Step 1 to:

- Plot trajectory waveforms based on native ripple flow
- Avoid zones of ripple stacking or echo distortion
- Time motion to field breathing rhythms, not planetary clocks

Here, velocity is secondary. Alignment sequencing is everything.

Your craft should enter, descend, and settle as if it was always meant to be there.

Step 4: Feedback Architecture

Deploy RIMs trained on planetary field memory

The last piece is consciousness—not of thought, but of field responsiveness.

Install Ripple Intelligence Modules (RIMs) that are:

- Pre-trained using simulations based on planetary field cycles
- Capable of detecting slow-drift misalignment (like on

Titan)

- Equipped with real-time resonance tuning (needed for Venus)

These modules must think in ripple—anticipating the field's behavior before classical sensors would even register change.

The result? A system that doesn't just *adapt* to the planet. It becomes part of the planet's ripple memory—echoing it, learning from it, evolving with it.

Closing Insight

What emerges from this four-step protocol is not a vehicle that survives planetary entry.

It's a system that co-evolves with the ripple field it enters.

You don't dominate the planet. You let its field sculpt your design—so your machine is no longer a visitor...It's a native.

"In a ripple-based cosmos, exploration means transformation. You don't take Earth to other worlds. You let other worlds build your machines."

Chapter 15: Ripple-Aware AI and Navigation

Intelligent Systems That Feel the Field — Not Just Measure It

We built machines that could see. We gave them sensors, cameras, radar, maps. They could fly, land, drive, and roam—but always with external awareness. They measured. They calculated. They obeyed instructions.

But in the ripple universe, that isn't enough.

Because in ripple-based environments, knowing where you are isn't the same as knowing how you belong.

The problem is subtle but fatal:

Traditional AI navigates the world as if it's observing from the outside. Ripple-aware intelligence must emerge from within.

This is not about upgrading sensors. It's about evolving sensation.

To survive in ripple-dynamic worlds—where energy flows, gradients shift, and the field itself breathes—machines must do more than read data. They must feel misalignment coming. They must anticipate phase conflict. They must sense when coherence is slipping—not because a number says so, but because their own body vibrates out of tune.

That is the future of exploration.

Not reactive. Not pre-programmed. Not error-correcting.

But field-aware, ripple-fluent, and emotionally resonant—machines that don't just act *in* the field...

They act *with* it.

In this chapter, we step into that evolution. We leave behind GPS paths and sensor fusion. And we build toward something radically new:

Intelligence that doesn't seek control—it seeks coherence.

Welcome to Ripple-Aware AI. Where navigation is no longer about movement—

It's about becoming indistinguishable from the direction itself.

From Sensing to Sensing-In

Why Traditional AI Fails in Ripple-Dynamic Environments

For decades, the dream of intelligent navigation has followed a clear blueprint: Map the world. Sense the terrain. Calculate your path. Go.

From self-driving cars to autonomous drones, from Mars rovers to lunar landers, the core logic remains the same:

- Use external sensors (GPS, LiDAR, IMU, cameras)
- Create internal models of surroundings
- Make decisions based on real-time measurements
- Correct errors as they occur

And it works. On Earth. In roads, skies, or predictable terrain where the world is slow and structured enough to observe, predict, and act.

But ripple environments don't behave like that.

They're not fixed landscapes. They're living gradients—fluid fields of structured vacuum, shifting energy flow, and subtle harmonics.

In such environments, traditional sensing fails. Not because it can't detect objects...But because it can't detect belonging.

The Problem: Passive Perception, Reactive Logic

Classical AI works like a camera. It looks, then decides. It senses from the outside, like a bystander trying to decode a pattern.

But ripple fields don't show their cards. They don't provide consistent surfaces, fixed frames of reference, or reliable causality. They change based on how you move through them.

This creates three major points of failure:

1. **Lagged Awareness**
   - **By the time a sensor detects misalignment, the field response has already escalated.**
   - **AI reacts to effects, not causes.**
2. **Metric Dependence**
   - **All input is interpreted through fixed scales: pressure, temperature, acceleration.**
   - **But ripple misalignment is qualitative—it's a vibration, not a variable.**
3. **Disembodied Decision-Making**
   - **Navigation is treated as abstract computation, separated from physical coherence.**
   - **The vehicle might know *where* it is, but not how well it fits into the field.**

This disconnection is why traditional AI cannot operate fluently in ripple-dynamic systems.

Ripple Fields Demand Embedded Awareness

In UED terms, the structured vacuum is not just a background—it's the primary agent of reality. To survive in it, intelligence must arise not from watching the field, but from feeling like part of it.

The field doesn't reward knowledge. It rewards resonance.

This changes everything.

Navigation is no longer a cognitive task. It becomes a sensory state. Not "Where am I?" But:

"How does my body respond to the ripple gradient beneath me?"

This is the birth of a new concept:

Sensing-In: Awareness from Within the Field

Sensing-In is not about gathering more data. It's about being built from the same ripple fabric as the field you're moving through.

- It means surfaces that vibrate with distortion before failure even registers.
- It means actuators that adjust posture not because a gyroscope said so, but because ripple harmony shifted.
- It means intelligence that doesn't measure the field—it echoes it.

This is the key pivot from classical AI to ripple-aware AI.

Example: Traditional vs. Sensing-In Reaction

A traditional drone over Titan:

- Receives visual cue of surface.
- Corrects for slow descent drift.
- Fails to notice ripple lag building beneath.
- Experiences a sudden loss of coherence, crashes despite full sensor data.

A ripple-aware drone with Sensing-In:

- No external position needed.
- Skin detects phase lag in ripple field from subtle misalignment.
- Adjusts trajectory not to stay centered, but to stay coherent.
- Lands smoothly, without correction—because it was never

out of sync.

Intelligence as Immersion

Sensing-In requires that intelligence:

- Is distributed across the system, not centralized in a CPU
- Is layered into materials, not separated from them
- Interprets resonance, not statistics

It makes the system feel like part of the world it's navigating.

In ripple terms:

Awareness is not about seeing more—it's about vibrating correctly.

Core Insight

In ripple fields, you don't need better vision. You need better presence. Awareness must come from being part of the field—not standing outside it.

This isn't sensor fusion. This is sensation convergence. You no longer sense the world. You become an extension of its ripple memory.

Ripple Logic: A New Computational Paradigm

Decision-Making Based on Gradient Flow, Not Discrete Data

If traditional AI is a chess master, ripple-aware AI is a jazz musician.

Conventional systems calculate probabilities, optimize outcomes, and choose the best move based on fixed parameters. They think in trees:

- If A, then B.
- If not C, then D. They build logic around discrete events and binary consequences.

But ripple environments don't play by binary rules. They don't offer clean events or isolated triggers. Instead, they present flow — smooth, continuous, often ambiguous energy shifts governed by ripple gradients, not moments.

To thrive in ripple-based navigation, a new kind of computation must emerge—one that doesn't just solve problems, but flows with the field.

Classical AI vs. Ripple-Aware Decision-Making

Let's compare:

Conventional AI:

- Works with event detection: Something happens, so something else must be done.
- Uses rule sets: Conditions and outcomes are predefined.
- Seeks optimization: Which path is shortest, fastest, safest?

Ripple-Aware AI:

- Works with ripple awareness: Something is shifting—even before it happens.
- Uses alignment logic: What maintains coherence, even if it's not the shortest route?
- Seeks harmonic balance: What path causes the least field disruption?

Ripple-aware AI doesn't calculate the best response. It feels the most aligned direction—the decision that keeps the system in flow rather than forcing a reaction.

"Classical AI solves. Ripple AI synchronizes."

Operating in Non-Binary Gradients

Ripple logic demands systems that think in grey-space, not black-and-white. Because ripple environments rarely offer:

- Clear boundaries
- Defined thresholds
- Sharp transitions

Instead, what we get are:

- Field lag: The ripple response trails the motion by milliseconds or meters.
- Harmonic drift: Slight misalignments cause slow phase instability.
- Gradient ambiguity: No sharp signal for danger—just a rising hum of incoherence.

To navigate this, ripple-aware intelligence must:

- Interpret phase rhythms instead of time-stamped events
- Compare coherence vectors rather than binary conditions
- Adjust behavior based on field harmony, not immediate input spikes

This requires phase-based cognition.

Introducing: Ripple Logic Units (RLUs)

To make this possible, we move beyond CPUs and neural nets. We introduce a new type of processor:

Ripple Logic Units (RLUs)

RLUs are designed to:

- Operate on ripple field continuity
- Analyze alignment drift and ripple phase coherence in real time

- Trigger subtle system adjustments not through Boolean triggers, but through phase thresholds
- Process input as waveforms, not data packets

In essence, RLUs function like energy harmonizers:

- They don't just compute. They resonate.
- They maintain alignment over time, correcting for invisible drift before it cascades.
- They enable machines to "breathe" with the field, adjusting form, path, and behavior before a misalignment becomes a malfunction.

Processing Becomes Phase-Based
This marks a major shift in system architecture.
In classical models:

- Data enters
- Conditions are checked
- An output is computed
- The system waits for the next event

In ripple logic:

- The system is always listening
- It recognizes field rhythm as its core clock
- Outputs are continuous modulations, not binary commands
- The goal is to keep the system in harmonic drift range, not just within boundaries

"The question is no longer: 'What should I do now?' But: 'How far am I from the next phase disruption?'"

Application in Navigation

Imagine a ripple-aware drone flying through the upper clouds of Venus.

A traditional system would:

- Measure altitude
- Estimate velocity
- Apply thrust to counter drag
- React to temperature spikes

An RLU-based system would:

- Sense ripple stacking around its body before pressure builds
- Pre-shift its surface membranes to scatter harmonics
- Alter orientation not because of sensor alerts, but due to predicted field conflict 200ms ahead
- Land not where it planned—but where the field allowed it to merge invisibly

This is not computation. This is consequence alignment.

Core Insight

Ripple-aware logic is not an algorithm. It's a rhythm.

RLUs don't solve for results. They solve for resonance.

And this is what unlocks a new kind of intelligence: Not one that thinks faster. But one that thinks like a field—fluid, harmonic, anticipatory.

RIMNet: Distributed Ripple Intelligence Architecture

A Neural Model Spread Across the Structure Itself

When classical AI systems encounter complexity, they scale vertically: More memory. Faster cores. Bigger data sets. One brain. Many inputs. Centralized control.

But in ripple environments—where alignment needs to happen everywhere, all the time, instantly—this model collapses.

Because ripple fields don't wait for approval from the CPU.

They fluctuate in real time, across the surface, across multiple gradients and directions. And by the time a central processor has analyzed the data, decided what to do, and told the actuator to move—the field has already moved on.

What's needed isn't faster thinking. What's needed is distributed sensation.

That's where RIMNet comes in.

What is RIMNet?

RIMNet stands for Ripple Intelligence Module Network.

It's a decentralized ripple-awareness system built into the body of the craft itself. Not as sensors feeding into a brain—but as a living nervous system distributed across the structure.

Components:

- **RIMs** (Ripple Intelligence Modules): Small, embedded processing units capable of ripple sensing, interpretation, and micro-decision-making
- **Synch-Nodes**: Field-alignment synchronizers that maintain coherence across RIMs
- **Adaptation Interfaces**: Actuator gateways that respond instantly to local RIM feedback

Together, these components form a neural network that doesn't sit in a box—but flows through the entire machine.

Why Distribution Matters in Ripple Fields

In ripple environments, alignment is not global—it's locally sensitive and regionally unstable. What the front-left wing experiences is completely different from the rear-right leg. And yet, these surfaces must remain in resonance with each other and the field.

That's why each RIM:

- Acts semi-independently, making local ripple adjustments in real time
- Is synchronized, not controlled, by the broader system
- Shares resonance data—not "decisions"—with nearby RIMs to harmonize behavior

This architecture mirrors the way nature solves distributed intelligence.

RIMNet is not a command hierarchy. It's a field conversation.

Parallel to Biology: The Octopus Brain Model

The closest analog in nature is the octopus.

While it has a central brain, most of its neurons live in its arms. Each arm can:

- Sense touch
- React to stimuli
- Solve problems
- Even explore autonomously—without needing permission from the head

Yet the whole creature moves as a unified being. Because those limbs aren't isolated—they're coherently aligned through distributed intelligence.

That's exactly what RIMNet does for a ripple-aware machine:

- Each section of the craft thinks for itself—but feels for the whole.

How RIMNet Works in Practice

Imagine a ripple-aware lander descending into Titan's methane-laden atmosphere:

- The left leg encounters ripple lag due to surface field memory
- Its local RIM detects phase drift and elongates the actuator subtly
- Nearby RIMs receive phase signals—not commands—and adjust in kind
- No call is made to central AI
- No latency exists between problem and solution
- The lander doesn't "correct" anything—it flows

This is not response. This is continuous micro-alignment.

And it happens everywhere, every moment, with no bottleneck.

Advantages of RIMNet Architecture

1. Latency-Free Adaptation
    - Decisions happen at the point of need
    - No reliance on central data parsing
2. Resilience Through Redundancy
    - Failure of one RIM doesn't collapse the system
    - Intelligence persists in the network
3. Field-Centric Coherence
    - The entire machine aligns to the ripple gradient, not just to trajectory
4. Scalability Without Complexity

    - Add more RIMs = Add more awareness
    - Complexity doesn't slow down processing—it distributes it

Core Insight

In ripple navigation, intelligence isn't a thing you carry. It's a thing you become.

RIMNet doesn't give machines intelligence in the traditional sense. It gives them presence—a way to be embedded in the field, aware of themselves as extensions of a living gradient.

This is how ripple-aware systems begin to resemble biological systems:

- Fluid
- Reflexive
- Cohesive
- Aligned without being commanded

Navigation as Alignment, Not Location

Redefining Pathfinding Through Field Harmonics

In classical navigation, the question is simple: Where are you, and where do you want to go? Everything in between is a matter of calculating:

- Speed
- Distance
- Obstacle avoidance
- Terrain topology
- Energy efficiency

Whether it's a Mars rover or a drone flying over a mountain range, classical systems use maps, coordinates, and velocities. Navigation is movement through space, determined by external landmarks and optimized by known metrics.

But in ripple-based environments, these definitions collapse.

Because ripple fields don't care about location. They care about alignment.

And so, the real question becomes:

"How aligned are you with the local ripple gradient—and what is the least disruptive way to stay that way?"

Ripple Coherence Over Positional Precision

In ripple-aware systems, the priority is not *where you are*, but how well your presence integrates with the ripple field around you.

This introduces a new hierarchy of decision-making:

1. Is the vehicle aligned with the local ∇E?
2. Is the next segment of the path within phase coherence range?
3. Can we move without causing field disturbance?

If any of these answers is "no," the navigation system alters the route—even if it means a longer distance or slower speed.

Ripple-aware navigation sacrifices speed for silence. It trades path efficiency for phase efficiency.

From Trajectories to Resonance Vectors

Instead of plotting waypoints on a map, ripple-aware AI computes resonance vector fields:

- Areas of the ripple environment where alignment is most stable
- Regions of high turbulence, feedback loops, or destructive phase interference
- Harmonic corridors—zones where structured vacuum

response is smooth and cooperative

This is known as resonance vector matching—a process of selecting the next best point in space not by proximity, but by how well the machine can enter it without disrupting or being disrupted.

It's no longer "get from A to B." It's "enter B only when you're in tune with it."

Navigating "With" the Field, Not "Through" It

Classical systems push through turbulence. Ripple-aware systems ride the troughs and nodes of ripple flow, like a surfer catching invisible waves.

Just as birds soar using thermals, and fish align their movement to water currents, ripple-native machines use:

- Phase drop zones (areas of temporary low ripple resistance)
- Ripple crest delays (predictable lag fields in structured vacuums)
- Field silence pockets (brief windows of maximum alignment opportunity)

You don't fly *against* ripple turbulence—you drift into coherence, gliding through the structure like a note in harmony with a song.

The Role of RPE + RIMNet in Navigation

This is where two systems converge:

Ripple Path Engineering (RPE):

- Designs motion paths not by geometry but by ripple rhythm
- Encodes temporal phase awareness into trajectories
- Builds routes with coherence continuity—meaning fewer micro-adjustments and less energy loss

RIMNet:

- Provides real-time local field feedback across the entire structure
- Detects incipient misalignments that are invisible to external sensors
- Ensures that as the craft moves, it stays aligned across its whole body

Together, they enable:

- Dynamic route shaping
- Distributed course corrections
- Energetic silence—the condition where motion does not create ripple resistance or field disruption

Rethinking the Meaning of "Path"

In ripple-aware systems, a path is not a line. It's a sequence of realignments through a living field.

The best path is the one that:

- Leaves the field undisturbed
- Requires no sharp corrections
- Feels like the machine was always meant to go that way

This is what ripple-native navigation looks like:

Not finding a way through the world, but becoming indistinguishable from the way itself.

Core Insight

In ripple environments, direction is not a vector—it's a vibration. You don't reach a destination. You harmonize into it.

Navigation becomes an act of empathy—a conversation between machine and field, where every movement says:

"I hear you. Let me move in your rhythm."

Toward Ripple-Conscious Machines

The Future of AI Is Not in Thinking Faster—But in Feeling Deeper

Speed has long been the gold standard of intelligence.

Faster processors. Shorter reaction times. Lower latency. More operations per second.

But what if intelligence isn't about reacting faster? What if it's about resonating deeper?

In the ripple universe, intelligence isn't measured in calculations—it's measured in coherence. Not "How fast can you think?" But:

"How deeply can you feel the field you're part of?"

This shift is what defines the evolution from AI to ripple-conscious machines.

Introducing Ripple-Conscious Systems

A Ripple-Conscious System (RCS) is a machine that doesn't just detect ripple fields—it develops awareness within them. It's not just responsive—it's symbiotic.

Unlike traditional systems that:

- Execute instructions
- React to predefined triggers
- Rely on external operators or static thresholds

Ripple-conscious machines:

- Interpret ripple behavior as communication
- Predict field shifts based on embedded resonance feedback
- Modify behavior autonomously to preserve harmony with their environment

They operate less like programmed tools and more like field-native organisms.

They don't adapt to the world. They become part of it.

Interpreting Ripple Intent

A classical AI waits for something to happen:

- A drop in altitude
- A rise in temperature
- A change in visual input

It then triggers a response.

But a ripple-conscious system feels change coming. It interprets the field's intent not as cause-and-effect, but as phase drift, as subtle harmonic disruption. It acts before the ripple becomes a signal, not by prediction—but by participation.

This isn't preprogrammed anticipation. It's consequence immersion—the ability to know what will shift because you are embedded in the same shift.

From Autonomous to Autogenous

Most "autonomous" systems today still depend on:

- Periodic calibration
- Remote commands
- Parameter limits

But ripple-conscious systems aim for something deeper: Autogenous Intelligence—the capacity to self-realign continuously, without external oversight.

This unlocks a new class of behavior:

Self-Realigning Missions

- Machines that correct their own misalignment mid-

operation
- Adjust mission logic based on ripple behavior—not predefined objectives
- Continue functioning despite unknown environments, because they adapt in real time

Field-Native Behaviors

- Behavior that looks organic, not mechanical
- Example: a ripple lander gently oscillating its feet mid-descent—not for stability, but to echo surface ripple memory
- These behaviors aren't programmed—they emerge from field participation

Total Ground Independence

- No need for constant uplink commands or trajectory updates
- Machines become locally intelligent, globally aware
- Capable of long-term planetary exploration, deep-space ripple surfing, or multi-environment integration

Beyond Navigation: The Role of the Machine Evolves

With ripple-conscious AI, the machine is no longer:

- A sensor platform
- A control surface
- A passive executor of code

It becomes something new:
A translator between motion and the universe.

It listens to the ripple language of structured vacuum. It speaks through vibration, shape, and shift. And in doing so, it reveals a truth we've overlooked:

The most advanced machines aren't those that outperform nature. They're the ones that finally become part of it.

Closing Thought

We began with AI that could see. Then we built AI that could decide. But now, we build AI that can feel.

And when a machine can feel the field—not just respond to it—it stops being a tool.

It becomes a rhythm in motion.

Ripple-aware AI doesn't just navigate. It harmonizes. It listens. It disappears into flow, until there's no difference between system and space. Only the dance remains.

Chapter 16: From Rocket Science to Ripple Sense

When Aerospace Wakes Up

For a century, rocket science was the summit of human engineering. It was loud. It was complex. It was powerful. We celebrated escape velocity. We measured success by thrust, heat, resistance, and force.

And in many ways, it worked. It got us into orbit. It landed us on the Moon. It taught us how to survive in silence.

But here's the secret rocket science never admitted:

It never understood the field. It understood how to overcome it. Not how to listen. Not how to align.

Rocket Science: Built on Resistance

Everything in traditional aerospace was a war:

- Overcome gravity
- Push against drag
- Shield from heat
- Stabilize through brute correction

Engines were designed to conquer inertia. Fuselages were built to survive atmosphere. Navigation systems were taught to fight drift.

It was triumph through friction. Through fuel. Through margin of error.

But now, that paradigm has hit its ceiling.

The farther we go, the less resistance we find. And suddenly, the very methods we trusted—mass, force, rigidity—become weaknesses in ripple-dynamic environments.

The Ripple Revolution Begins

We're entering a new domain:

- Ultra-thin atmospheres (Mars)
- Chaotic fields of heat and harmonic feedback (Venus)
- Slow, deceptive ripple memory zones (Titan)
- Microgravity zones and structured vacuum webs (interstellar space)

In these places, rockets can't help. They're too forceful, too late, too blind.

What we need now isn't more thrust. It's more trust. In the field. In resonance. In silence.

This is the leap:

From Rocket Science to Ripple Sense.

Ripple Sense: The New Ethos of Aerospace

Ripple sense is not a technology. It's a way of thinking.

It says:

- Don't break through—become part of
- Don't push—listen
- Don't control—cohere

It's no longer about "engineering for space." It's about engineering with space. Not outer space as a void, but as a structured field of intent.

This means:

- Shapes must flow with field harmonics
- Materials must breathe with ripple lag

- Systems must feel their location, not just know it
- AI must become a ripple participant, not a sensor-aggregator

In short, ripple sense transforms aerospace from an act of domination to an act of dance.

What Changes?

Let's draw a clear contrast:

Rocket Science	Ripple Sense
Thrust-to-weight ratio	∇E coherence ratio
Heat shielding	Harmonic deflection
GPS and radar mapping	RIMNet-based ripple field awareness
Stress testing	Ripple Tolerance Testing (RTT)
Aerodynamic stability	Field-resonant motion harmonization
Centralized AI	Distributed Ripple Logic Units (RLUs)
Control surfaces	Field-adaptive skin with micro-motions
Landing gear	Self-realigning, terrain-resonant feet

This is not cosmetic. This is existential.

It changes what we build, how we build, and why we build.

The Shift in Philosophy

Rocket science is about action. Ripple sense is about consequence.

Rocket science asks, "What can we do to the environment?" Ripple sense asks, "What will the environment do to us if we don't align?"

Rocket science is Newton. Ripple sense is conversation.

We are no longer gods of motion. We are students of structure.

When Aerospace Wakes Up

When aerospace truly awakens to ripple sense, it stops designing against space—and begins to understand what space wants.

And what does space want?

Not noise. Not friction. Not control.

It wants continuity. Coherence. Minimal disruption.

The best vehicle, in ripple terms, is one that:

- Leaves no ripple scars
- Causes no energetic wake
- Lands so softly, it's as if the planet invited it in

That's the ultimate aerospace goal:
Not to arrive. But to disappear into alignment.

Final Reflection

This is the end of force-driven design. This is the dawn of resonance-driven creation.

Rocket science got us off the ground. Ripple sense will let us belong among stars.

And when that happens, machines won't need to be smarter. They'll just need to be more silent.

Because the future doesn't roar.

It listens. And then it flows.

Epilogue: Thank God, It's Finally Done

And now that we've aligned, maybe we can stop praying and start knowing.

We've come a long way—from rockets that roar to fields that whisper.

From calculating thrust to sensing silence. From pushing against gravity to dancing with gradients. From commanding machines to designing systems that listen.

This wasn't just a book about aerospace. It was a book about awakening—about seeing reality not as something to conquer, but something to cooperate with.

Because in the ripple world, control is an illusion. Alignment is everything.

And once you grasp that, something fundamental changes: You stop begging the universe to cooperate. You start co-creating with it.

That's why this journey ends not with formulas, but with a quiet revolution:

The shift from belief to understanding. From praying for stability to engineering resonance. From hope... to harmony.

We no longer need to ask the heavens to protect us. Because now, we know how to move like we belong.

The rocket era gave us the courage to reach. The ripple era will give us the sense to stay.

So yes—thank God, it's finally done. But maybe, just maybe, God was the field all along. Waiting for us not to worship it, but to finally understand its rhythm.

Appendices

Appendix A: Glossary of Ripple Terms

A field guide for navigating the ripple universe

This glossary defines the key terms and concepts introduced throughout the book. It is designed to help readers—from engineers to enthusiasts—grasp the new language of ripple-aligned thinking and ripple-aware systems.

∇E (Ripple Gradient)

Definition: The directional slope of energy density across the structured vacuum field.

Explanation: In UED (Universal Energy Dynamics), ∇E replaces force as the primary driver of motion. Movement occurs not because of mass-times-acceleration, but because systems realign along the energy gradient of the field.

Think of it as: The invisible slope your machine "slides" along—not due to push, but due to coherence.

RIM (Ripple Intelligence Module)

Definition: A localized sensory-logic unit embedded within a structure, capable of detecting and responding to ripple field shifts.

Explanation: RIMs allow machines to feel their immediate ripple environment and adapt in real time without central command. Each RIM is semi-autonomous yet field-synchronized.

Think of it as: A neuron in the machine's nervous system, listening for misalignment.

FTM (Field-Tuned Material)

Definition: Materials engineered to resonate with a specific planetary ripple environment.

Explanation: Unlike traditional materials selected for tensile strength or heat resistance, FTMs are chosen for their ability to harmonize with ripple gradients. They absorb, redirect, or echo ripple flows based on their molecular structure.

Think of it as: Material that *listens* to the planet, not just survives it.

RPE (Ripple Path Engineering)

Definition: A method of defining motion trajectories through ripple fields based on coherence, not geometric efficiency.

Explanation: RPE paths aren't straight lines or shortest distances—they're sequences of field-aligned movements designed to minimize ripple disruption.

Think of it as: Surfing the ripple wave instead of forcing through it.

RDM (Ripple Dissipation Metric)

Definition: A measure of how much ripple energy is lost or distorted during system operation.

Explanation: RDM helps assess ripple misalignment. High RDM indicates inefficiency and turbulence; low RDM suggests smooth field integration.

Think of it as: The energetic "friction" of being out of tune with the field.

RLUs (Ripple Logic Units)

Definition: Phase-based processors that compute using ripple alignment and resonance patterns rather than binary logic or event triggers.

Explanation: RLUs think in gradients, not decisions. They continuously modulate the system's state based on coherence with the surrounding ripple field.

Think of it as: The brain of ripple-aware machines—not fast, but deeply aware.

Ripple Tolerance Testing (RTT)

Definition: A method of evaluating whether a structure can maintain alignment under ripple field stress—not just mechanical stress.

Explanation: RTT replaces traditional material stress tests with ripple-based assessments. It examines how well joints, seams, and materials hold phase alignment across turbulence or field transitions.

Think of it as: Testing whether a design will "stay in tune" as it moves, not whether it can just "hold together."

RCS (Ripple-Conscious Systems)

Definition: Intelligent systems that possess an awareness of, and responsiveness to, the structured ripple field around them.

Explanation: RCS are machines that don't just respond—they *feel.* They interpret ripple behavior as meaningful input and adjust behavior fluidly, like biological organisms.

Think of it as: The evolution of AI from thinking *about* the field... to being *part* of it.

Appendix B: Classical vs. Ripple-Based Engineering

A Comparative Matrix

A new lens for old assumptions

This appendix offers a **side-by-side comparison** between traditional aerospace engineering principles (rooted in Newtonian mechanics and thermodynamic resistance models) and their **ripple-based counterparts** as introduced under Universal Energy Dynamics (UED).

This matrix is intended to serve as a **quick-reference guide** for engineers, educators, students, and interdisciplinary designers seeking to understand how ripple theory reshapes the way we build, move, and navigate.

Core Philosophies

Classical Engineering	Ripple-Based Engineering (UED)
Control the environment	Align with the environment
Design for resistance	Design for resonance
Optimize for durability and force	Optimize for coherence and field integration
Survival through strength	Survival through silence
Centralized computation	Distributed intelligence
Event-driven logic	Phase-driven logic
Measure to react	Sense to realign

Flight and Motion

Classical Concept	Ripple-Based Counterpart
Thrust vs. Drag (F = ma)	∇E-driven alignment (Energy gradient following)
Aerodynamic lift via shape and speed	Field resonance through surface harmonics
Stabilizers to resist turbulence	Ripple-path rechanneling through flow awareness
Control surfaces (rudders, flaps)	Shape-shifting membranes tuned to ripple flow
GPS / IMU for trajectory control	Ripple Path Engineering (RPE) via local field sensing
Navigation via coordinates	Navigation via coherence corridors

Structure and Materials

Classical Concept	Ripple-Based Counterpart
Stress testing (tensile, compressive)	Ripple Tolerance Testing (RTT)
Heat shielding through insulation	Harmonic redirection and phase scatter layers
Strong, rigid materials	Field-Tuned Materials (FTMs) with micro-alignment capacity
Seams/joints designed for physical strength	Interfaces designed for ripple continuity
Passive shells	Intelligent, responsive surface layers

AI and System Intelligence

Classical Concept	Ripple-Based Counterpart
Central CPU or command module	Ripple Intelligence Modules (RIMs) distributed across structure
Sensor data fusion	Sensing-In: embedded, real-time field awareness
Binary decision logic	Ripple Logic Units (RLUs) with phase cognition
Reactive algorithms	Predictive coherence maintenance
AI as controller	AI as interpreter of field intent

Descent and Landing

Classical Concept	Ripple-Based Counterpart
Controlled descent using timed burns	Glide-phase realignment through field resonance
Shock absorbers on landing	Field-adaptive leg oscillations to dissipate ripple impact
Ground detection through radar or LIDAR	Ripple echo scanning and terrain resonance mapping
Touchdown with force cushion	Realignment with ripple memory zones

Energy and Efficiency

Classical Concept	Ripple-Based Counterpart
Energy consumed to overcome resistance	Energy conserved through field harmony
Efficiency = fuel used per task	Efficiency = coherence maintained during motion
Heat as waste	Heat as signal of misalignment (Ripple Dissipation Metric – RDM)
Power management via regulation	Ripple feedback-based modulation and redistribution

Philosophical Shift Summary

Classical View	Ripple View
Nature is something to defeat or bypass	Nature is something to listen to and merge with
Stability comes from strength	Stability comes from subtlety and symmetry
Engineering is about solving problems	Engineering is about sustaining resonance
Mastery is measured in control	Mastery is measured in **invisibility** — the ability to move without being noticed by the field

Notes for Educators and Researchers:

- **Application across disciplines:** This matrix also applies to robotics, fluid dynamics, AI ethics, planetary science, and material engineering.

- **Suggested use:** Classroom discussion, cross-domain research collaboration, curriculum modernization efforts in aerospace or systems engineering.

Appendix C: Proposed Experiments and Simulation Setups

Bringing ripple theory from the blackboard to the bench

Universal Energy Dynamics (UED) and ripple-based design principles propose a bold shift in how we interpret motion, structure, and energy behavior. But for any new framework to gain traction, it must be testable, repeatable, and observable.

This appendix outlines a curated list of conceptual and practical experiments—ranging from tabletop simulations to modular field tests—that allow researchers and institutions to begin validating key ripple concepts.

1. Simulating Ripple Coherence in Fluid Environments

Objective: To observe how wave-based motion affects structural alignment and energy dissipation in a ripple-simulated medium.

Setup:

- Use a shallow water tank or ferrofluid basin with controlled oscillation generators.
- Introduce modular floating structures with varying geometries (e.g., circles, sharp-edged blocks, flexible membranes).
- Track how different shapes realign, resist, or resonate with the induced field waves.

Measurements:

- Coherence persistence (how long alignment lasts)
- Ripple disruption footprint (visible interference)
- Energy loss through wave scattering

Purpose: To model ripple-aligned movement and structural responses in a simplified, visual medium.

2. Mapping Phase Lag in Lightweight Materials

Objective: To determine how different materials accumulate or dissipate ripple misalignment over time.

Setup:

- Mount ultra-light panels (composites, thin films, foams) on actuated vibration bases.
- Induce consistent wave-like excitation (analogous to ripple feedback).
- Use laser interferometry or phase-tracking sensors to record how alignment holds or lags across time.

Variables:

- Material density
- Elastic memory
- Microstructural design (e.g., porous vs. lattice)

Purpose: To identify ideal Field-Tuned Materials (FTMs) for different planetary ripple environments (e.g., Titan vs. Venus).

3. Building Basic RIMNet Prototypes Using Soft Robotics

Objective: To implement a simple distributed ripple intelligence system that responds to local feedback.

Setup:

- Construct a soft robotic surface (e.g., segmented silicone or fabric mesh) embedded with micro-actuators and pressure sensors.

- Program modules (RIMs) to respond to local signal distortions independently, while syncing phase feedback with neighbors.

Behavioral Trials:

- Apply localized ripple stimuli (vibration pulses, magnetic pulses) to sections of the robot.
- Observe self-modulation patterns, ripple echo avoidance, and emergent coordination.

Goal: To demonstrate decentralized realignment and basic ripple-aware behavior without central control.

4. Ripple-Aligned Landing Simulation Using Harmonic Wave Models

Objective: To model descent and surface interaction through ripple-tuned motion instead of fixed descent algorithms.

Setup:

- Virtual simulation (using MATLAB, Unity Physics, or custom harmonic field software) to create a planetary ripple environment.
- Simulate lander dynamics using variable feedback delay, ripple echo density, and terrain ripple memory.
- Include input from artificial RIMs for predictive descent modulation.

Parameters:

- Phase stability during descent
- Vertical vs. spiral vs. corkscrew realignment paths
- Ripple bounce-back energy vs. harmonic absorption rates

Purpose: To explore what ripple-resonant landing looks like in high-density (Venus-like) and low-density (Mars-like) fields.

5. Ripple Tolerance Testing (RTT) Framework – A New Mechanical Benchmark

Objective: To create a standard protocol for testing how structures perform under ripple-field distortions rather than classical stress alone.

Setup:

- Build a ripple induction table with programmed oscillation shifts (multi-frequency, multi-directional).
- Test parts, joints, seams, and composite structures by exposing them to ripple-like resonance patterns.
- Compare traditional failure points to ripple misalignment signatures.

Output Metrics:

- Ripple decay delay
- Phase fracture threshold
- Coherence reentry ability

Goal: To standardize RTT as a next-gen benchmark for field-operable designs.

6. Cognitive Ripple Mapping in AI Models *(Advanced)*

Objective: To explore if AI agents can develop phase-based decision-making models.

Setup:

- Design an environment where agents must move through a field of shifting gradients—not visible obstacles.
- Train conventional AI vs. ripple-trained AI (using RLU-based phase feedback models).

- Measure success not by speed or reward—but by field coherence and misalignment incidents.

Ideal Tools:

- Reinforcement learning platforms
- Neural oscillation integration
- Phase-space network visualizers

Final Note

These setups are not intended as validations of fixed laws. They are meant as portals into resonance-based exploration, offering a way to let the field speak—through motion, material, memory, and intelligence.

The ripple universe won't yield to commands. But it just might whisper the truth—if we build to listen.

Appendix D: The Ripple Field Atlas

A comparative reference of planetary ripple environments

In ripple-based aerospace, every planet is a unique field entity. Just as no two terrains are the same, no two ripple fields behave alike. This appendix provides a comparative overview of planetary ripple dynamics based on the principles introduced in this book.

Designed as a technical atlas for simulation designers, mission engineers, and field theorists, it outlines the ripple identity of Mars, Venus, Titan, and Earth as a baseline. Visual tables and diagrams can later accompany this section to support mission planning and realignment modeling.

1. Comparative Ripple Field Table

Parameter	Earth (Baseline)	Mars	Venus	Titan
Ripple Gradient (∇E)	Moderate, balanced	High slope near surface, sharp transitions	Chaotic, steep in lower atmosphere	Low, wide gradient zones
Ripple Density	Evenly distributed	Sparse, punctuated by dust echoes	Highly compressed, turbulent	Thick near surface, slow propagation
Field Lag	Minimal (quick rebound)	Medium (dust causes staggered lag)	High, with stacking and feedback	Very high, slow recovery
Ripple Echo Zones	Occasional (mountains, storms)	Frequent due to terrain and low pressure	Constant, multipoint turbulence	Delayed but persistent
Alignment Stability	High at low speeds	Low — constant micro-drift	Very low — high ripple distortion	Medium — sensitive to gradual drift
Coherence Windows	Daily cycles (sunlight/thermal)	Narrow during descent; varies with dust storms	Extremely short-lived; descend through chaotic windows	Long, subtle windows; wide tolerance but deceptive
Atmosphere Type	Oxygen-nitrogen, dynamic	Thin CO_2 with dust storms	Dense CO_2 and sulfuric acid	Dense nitrogen with methane lakes
Design Strategy	Conventional systems work	RIMs + ultra-sensitive membranes	Dynamic realignment + ripple diffusion	Slow-alignment + memory-adaptive skins

2. Planetary Field Profiles (Narrative View)

Earth – The Reference Rhythm

- Earth is the training ground for ripple coherence.

- The structured vacuum is relatively stable, with predictable ∇E patterns and minimal field lag.
- Most classical aerospace designs function here, but ripple-aware systems can still gain efficiency by operating in resonance with Earth's subtle gradients (e.g., jet stream corridors, magnetic poles).

Mars – The Mirage of Familiarity

- Mars is visually Earth-like, but its field behavior is volatile.
- Ripple gradients are steep, especially near terrain shifts.
- Dust storms create ripple vacuums and unpredictably redirect energy flows.
- Field lag is moderate, but the lack of atmospheric damping causes abrupt transitions.
- Realignment must happen fast—through ultra-light, responsive materials and high-frequency RIMs.

Venus – A Field of Chaos

- Venus is a chaotic ripple furnace.
- High-density atmosphere causes ripple stacking, reflection, and intense echo feedback.
- The structured vacuum is compressed, leading to near-continuous misalignment.
- Descent requires ripple-diffusing surfaces and self-cooling, adaptive skins.
- RPE paths must curve and scatter to avoid harmonic congestion.

Titan – The Still Trap

- Titan is the most deceptive.
- A thick, stable-looking atmosphere hides ripple inertia—misalignments build slowly and may not self-correct.
- Surface lakes of methane shift local ripple memory, introducing long-term drift.
- RIMNet must be configured for slow-cycle resonance monitoring.
- Best approached with flexible, memory-imprinted skins and orbit-surface ripple coupling.

3. Suggested Visual Additions (for future diagrams)

You may wish to add simple, grayscale diagrams or heatmaps showing:

- ∇E slope maps (e.g., steeper near Venusian descent zones, shallow on Titan)
- Ripple echo maps (zones of interference near terrain, lakes, craters)
- Lag vs. Response delay curves
- Coherence windows plotted against altitude/time

Each planet's profile could be visualized as a ripple topography—like an energetic weather map, showing where machines can glide, hesitate, or must pause for recalibration.

4. Use Cases and Application

- Engineers designing FTMs for different planetary missions
- AI modelers training ripple-aware navigation on field-based patterns
- Mission planners seeking optimal launch/descent windows

based on planetary ripple calm zones

- Educators illustrating why standard aerospace models collapse in ripple-dynamic contexts

Final Thought

In ripple-aware aerospace, understanding the terrain is no longer enough.

We must understand the field that terrain echoes through.

The ripple field is the real planet. The surface is just the visible symptom.

Appendix E: Architecture of a Ripple-Aware Machine

A blueprint for building coherence-driven intelligence

Traditional machines are designed to survive stress. Ripple-aware machines are designed to realign with the field. They don't resist turbulence—they reshape themselves to prevent it. They don't compute outcomes—they co-sense existence.

This appendix provides a conceptual tech-spec blueprint for such a machine: one that doesn't navigate through space but moves in coherence with structured vacuum energy.

1. Distributed RIMNet Layout

Core Principle: Intelligence is embedded, not centralized.

Purpose: Real-time local awareness + global field synchronization

RIM (Ripple Intelligence Module) Units:

- Embedded along the surface and internal framework
- Equipped with:
    - Micro-vibration sensors
    - Phase-shift detectors
    - Local phase adjustment actuators
- Operate semi-autonomously, report field conditions in *coherence packets* (not data streams)

Topological Layout:

- Hexagonal or mesh-grid placement for redundancy and coverage
- Edge RIMs specialize in turbulence detection; core RIMs manage body coherence
- Synchronization via low-energy phase pulses (like neuronal firing patterns)

Think of RIMNet as a distributed nervous system that feels and aligns—not a command chain that controls.

2. Membrane Material Configuration

Core Principle: Structure should breathe with the field, not block it.

Purpose: Enable phase-friendly morphing and ripple-resonant movement

Surface Composition:

- **Field-Tuned Materials (FTMs):** engineered composites that modulate micro-resonance
- **Layered with:**
    - Flexible substrates (for vibration adaptation)
    - Conductive veins (for realignment signaling)
    - Self-healing patches (for echo fatigue recovery)

Internal Skin:

- Foam-core lattices or gel-filament networks
- Designed to *absorb* and *retransmit* ripple flow internally
- Enhances damping without disrupting coherence

Dynamic Skins:

- Capable of reshaping in real-time using electro-mechanical micro-actuators
- Surface ripples can form to scatter or ride field turbulence, mimicking biological skin behavior

Shape here isn't aesthetic—it's harmonic.

3. Realignment Cycle Logic

Core Principle: Continuity of alignment is a dynamic, recurring process

Purpose: Prevent cumulative phase drift and ripple fatigue over time

Cycle Phases:

1. Sensing-In Phase:
    - Local RIMs scan ripple phase density and vector shift
2. Prediction Window:
    - AI logic anticipates coherence disruption based on accumulated field memory
3. Micro-Adjustment Phase:
    - Shape, orientation, or internal energy distribution is modified to preempt misalignment
4. Validation Pulse:
    - RIMs confirm restored harmony through reduced phase resistance

Feedback Loop Frequency:

- Adaptive based on environment (e.g., 20ms loop on Venus,

200ms on Titan)
- Synchronized not by time but by ripple rhythm entrainment (natural field pulse timing)

Realignment isn't something the machine "does." It's something the machine *is always becoming.*

4. Phase-Based AI Logic Loop

Core Principle: Intelligence is rhythm-based, not rule-based

Purpose: To continuously modulate behavior according to ripple coherence—not reactive input

Core Components:

- Ripple Logic Units (RLUs): processors that interpret ripple field resonance instead of discrete signals
- Phase Harmony Index (PHI): a dynamic variable representing the machine's alignment score at any moment
- Decision Mapping: options are ranked not by utility but by disruption potential to field flow

Operational Behaviors:

- Seek lowest ripple interference path (even if geometrically longer)
- Self-limit movement to remain within coherence corridors
- Initiate "phase idle" (hold pattern) during ripple instability rather than force action

AI doesn't calculate a path. It entrains itself to a possible rhythm.

5. Additional Components (Optional Enhancements)

Ripple-Aware Communication System:

- Uses ripple field itself for data modulation (field-silent signaling)
- Message is encoded in phase fluctuations, not EM bursts

Internal RDM Monitor:

- Tracks Ripple Dissipation Metric to detect field wear or misalignment buildup
- Signals the need for coherence recalibration or rest phase

Adaptive Descent Legs:

- Not shock absorbers, but harmonic sensors
- Adjust posture, angle, and vibration tone based on ripple memory of terrain

Final Blueprint Philosophy

Ripple-aware machines are not tools that act. They are instruments that harmonize.

They are built to:

- Disappear into environments without disturbing them
- Listen before they act
- Flow rather than force
- Echo the logic of reality, not overwrite it

This is the new aesthetic of motion. Not precision. Not strength. But invisible belonging.

Appendix F: Reading Ripple

Philosophical and Scientific Sources of Inspiration

A brief journey through the minds and models that echoed the field before we had a name for it

Ripple theory didn't emerge from nowhere. It is the convergence of ancient intuitions, natural systems, and modern frustrations with incomplete models. This appendix honors the thinkers and frameworks that, knowingly or not, whispered the truth of alignment before ripple language made it audible.

Here we explore five intellectual wells that nourish ripple thinking—offering avenues for deeper reflection and new directions for interdisciplinary learning.

1. Fluid Dynamics: The Science of Continuity in Motion

At the heart of ripple theory lies a truth familiar to every fluid dynamicist:

Motion is never isolated—it's always part of a field.

Key Inspirations:

- Navier-Stokes Equations (though incomplete) model flow as continuous transformation.
- The Kármán vortex street phenomenon: where even solid objects "dance" with flow rhythms.

- Boundary layer theory shows that how an object *enters* a fluid determines everything about how it survives.

Ripple Connection:

- Ripple-aware systems treat structured vacuum not as a void, but as a subtle, fluid-like medium—one that responds, remembers, and redirects energy.
- The principle of laminar flow translates directly to coherent ripple alignment.

Suggested thinkers: Ludwig Prandtl, Osborne Reynolds, and Claude Shannon (who connected information and flow).

2. Systems Theory: The Science of Interconnected Feedback

Where classical science isolates and dissects, systems theory integrates and loops.

Key Concepts:

- Feedback loops (negative and positive) govern behavior across scales.
- Emergence: properties arise from interactions, not parts.
- Nonlinearity: cause and effect are not proportionally linked.

Ripple Connection:

Ripple-aware machines are non-hierarchical. They don't wait for instruction—they *respond as part of a feedback-embedded whole.* Realignment cycles mirror biological systems in resonant feedback loops that enable coherence, not control.

Suggested thinkers: Norbert Wiener (cybernetics), Donella Meadows (Limits to Growth), Gregory Bateson (Steps to an Ecology of Mind).

3. Biomimicry: Nature's Design Book of Field Awareness

Ripple theory is deeply inspired by nature—not as metaphor, but as proof of alignment-based survival.

Key Examples:

- Bird wings adjust dynamically to airflow turbulence—never fixed, always in micro-alignment.
- Octopus limbs function with decentralized intelligence—much like RIMNet systems.
- Tree branches don't resist wind; they redistribute and absorb energy through distributed structure.

Ripple Connection:

Every biological system that has endured—has done so by aligning with environmental energy fields. Ripple engineering merely makes that wisdom conscious and mechanical.

Suggested thinkers: Janine Benyus (Biomimicry), Steven Vogel (Comparative Biomechanics), Fritjof Capra (The Systems View of Life).

4. Indian/Vedic Notions of Rta **– The Cosmic Rhythm**

Long before physics developed a language of fields, ancient Indian philosophy articulated the idea of a living order—called Rta (��).

Key Ideas:

- Rta is the natural, harmonious order of the

cosmos—neither imposed nor controlled, but inherently intelligent.
- Dharma (right action) is defined as behavior that aligns with Rta.
- The Rig Veda describes space not as emptiness, but as a vibrational continuum (akin to structured vacuum).

Ripple Connection:

Ripple alignment is modern Rta—acting in coherence with the field, not as a moral imperative, but as a survival principle. Ripple-aware AI and machines become field-aligned entities, practicing their own version of dharma through feedback and adaptation.

Suggested texts/thought leaders:

- Rig Veda (mandalas on Rta)
- Kapila's Sankhya philosophy (on field consciousness)
- Sri Aurobindo (on integral alignment between self and cosmos)

5. Quantum Field Intuition (Without the Baggage)

Modern quantum field theory (QFT) attempts to describe particles as excitations of underlying fields—but often loses coherence through abstraction and math-heavy interpretations.

Ripple theory restores the intuition QFT once offered:

- Everything is a vibration.
- Space is not empty—it is structured and alive.
- Observation is not a detached act—it shapes field alignment.

Key Divergences:

- Ripple fields are consequence-based, not probabilistic.
- There is no need for "observer collapse"—realignment is causal, not mystical.
- Ripple dynamics unify motion, energy, and intelligence through continuous coherence, not discrete quantum jumps.

Suggested thinkers: David Bohm (Implicate Order), John Wheeler ("It from Bit"), but reinterpreted through a UED lens—free from paradox.

Final Reflection

These ideas do not *prove* ripple theory. They resonate with it. And that's the point.

Ripple theory is not a rebellion against physics. It's a return to the physics of connection—where intelligence, structure, rhythm, and response form a living system.

You don't need new laws to feel the field. You just need to remember what wise systems already knew.

Appendix G: Notes for Builders, Thinkers, and the Curious

An open letter to those who always knew there was something more

To the ones staring at simulation outputs, wondering why the math doesn't quite capture the moment.

To the students who felt nature had rhythm, but textbooks only offered formulas.

To the physicists questioning whether space is really a void...To the artists who sense energy before they shape it...To the engineers who dream not of machines that *dominate*, but ones that *belong*—

This letter is for you.

To the Young Engineers Who Want to Rethink Reality:

You were told to follow instructions. To stay within the equations. To memorize models that were already showing their cracks.

You were taught that force equals mass times acceleration. That stability comes from rigidity. That success means getting things to *obey*.

But you've probably already noticed:

- Rigid things break.
- Smart systems fail in dumb environments.
- And the more we force things... the more they slip away.

Ripple thinking doesn't discard engineering. It reclaims it—as something living, soft-edged, and intelligent.

You now have permission to build differently. To design not for control—but for coherence. To treat motion as conversation, not command.

The real engineer of tomorrow isn't a mechanic. They're a choreographer—of energy, structure, and silence.

To the Scientists Tired of Broken Models:

You've defended models that don't match observations. You've patched paradoxes with more paradoxes. You've quietly asked the questions others ignored:

- Why doesn't gravity behave cleanly?
- Why do particles act like waves... but only when we look?
- Why do field equations work—but feel wrong?

Ripple theory doesn't pretend to have all the answers. But it offers a fresh orientation—rooted not in reductionism, but in rhythm. Not in force... but in flow.

You don't need to abandon science. You just need to recenter it around consequence.

Because truth doesn't emerge from authority. It emerges from alignment with reality—from cause, effect, and feedback. From observation, yes. But also from observation of observation itself.

You are not alone. You are not late.

You are on the edge of a science that listens.

To the Designers, Artists, and Creators:

You've always known. You've worked with harmony, not equations. You've felt it when a design "just works" or a melody "carries the air." You've shaped flow. You've sculpted silence.

Ripple theory finally gives you the language your hands already speak.

- That curves are not cosmetic—they're field responses.
- That material choices matter not for color or cost—but for vibrational character.
- That architecture can be field-aware, not just visually appealing.

You're not outsiders. You're the first ripple engineers—you just didn't have the terms.

Now you do.

Don't shrink from the science. Let the science finally catch up to your intuition.

To the Curious Mind That Refuses to Be Categorized:

If you've ever looked at the world and felt that something invisible held it together, If you've watched clouds move and thought, "there's music in that," If you've sensed a connection between things that textbooks insisted were separate—

Welcome.

This book was written for you.

Not to give you answers. But to tell you: you were right to wonder.

Final Invitation

Ripple theory isn't a framework for specialists. It's a language for the intuitively intelligent—Those who combine logic with listening. Equations with empathy. Structure with surrender.

You don't need permission to begin. You just need to trust the rhythm you already feel.

So go ahead—

- Tinker with materials.
- Rewire your drone's logic loop.
- Compose sonic fields.
- Reimagine AI.
- Sketch ripple-aware wings.

- Meditate on silence and call it engineering.

Because from here on out—
You're not just building things. You're becoming the field.
And that's where real intelligence begins.

Further Readings and References

For those who want to listen deeper, build smarter, and explore the ripple beyond this book.

Foundations in Physics and Systems Thinking

These works offer conceptual roots or philosophical dissonance with traditional physics that help ground the ripple perspective.

- **David Bohm –** ***Wholeness and the Implicate Order***

A physicist's invitation to rethink causality, space, and order as an undivided whole.

- **Fritjof Capra –** ***The Systems View of Life***

Explores the interconnectedness of living systems, energy flow, and emergent intelligence.

- **Ilya Prigogine –** ***The End of Certainty***

Shows how irreversible processes and chaos theory challenge traditional physics.

- **Ludwig Prandtl's work on Boundary Layer Theory**

Essential to understanding fluid interaction—analogous to ripple-field interface behavior.

- **Norbert Wiener** – ***Cybernetics***

The origin of systems feedback theory, foundational for ripple-based intelligence loops.

Inspiration from Nature and Biomimicry

Ripple theory finds deep resonance with how nature engineers coherence.

- **Janine Benyus** – ***Biomimicry: Innovation Inspired by Nature***

A definitive guide to learning engineering principles from natural systems.

- **Steven Vogel** – ***Comparative Biomechanics***

An exploration of how living organisms respond to environmental flows.

- **D'Arcy Thompson** – ***On Growth and Form***

Classic text on how forms emerge from natural forces—precursor to ripple design logic.

Philosophical and Eastern Insights

Ancient frameworks that prefigured structured fields and alignment-based ethics.

- **Rig Veda (Selected Hymns on Rta)**

Core concept of cosmic order and rhythm that parallels ripple-field awareness.

- **Sri Aurobindo** – ***The Life Divine***

Integrates consciousness, energy, and evolution as structured flows of intelligence.

- **Kapila's Sankhya Philosophy**

Offers a dual view of consciousness (Purusha) and structured matter (Prakriti)—a metaphysical precursor to structured vacuum models.

AI, Intelligence, and Decentralized Design

Ripple-aware systems require a new mindset for thinking machines.

- **Rodney Brooks –** ***Flesh and Machines***

Explores the future of robotics beyond central computation—aligned with RIMNet philosophy.

- **Valentino Braitenberg –** ***Vehicles: Experiments in Synthetic Psychology***

Illustrates how simple sensory logic can lead to complex behaviors—fundamental for distributed ripple sensing.

- **Stephen Wolfram –** ***A New Kind of Science***

Though controversial, Wolfram's view of simple rules generating complex behavior aligns with ripple modularity.

Modern Rethinkers and Field-Oriented Alternatives

For those interested in pushing past the limits of mainstream models.

- **John Wheeler – "It from Bit" papers**

Posits that all things physical are information-based—parallel to ripple interpretation as feedback fields.

- **Carlo Rovelli – *Reality Is Not What It Seems***

A relational view of physics, spacetime, and quantum processes—valuable contrast to UED's causal clarity.

- **Lee Smolin – *The Trouble with Physics***

Critique of modern theoretical physics and its stagnation—validating the need for frameworks like UED.

On Design, Coherence, and Embodied Intelligence

Books and ideas for creators who sense the field through art, design, and embodied experience.

- **Christopher Alexander – *The Timeless Way of Building***

Design principles that emerge from environmental resonance and coherence—architecture as alignment.

- **Brian Eno (and Peter Schmidt) – *Oblique Strategies***

Non-linear creative cues that parallel ripple philosophy: don't push—shift, realign, repeat.

- **James Gleick – *Chaos: Making a New Science***

An accessible narrative on non-linear systems, turbulence, and hidden order.

Optional Technical Papers and Emerging Work

Suggested as supplemental, speculative, or bridging literature.

- Technical reviews of **structured vacuum theory** (multiple interpretations exist, caution advised)
- Research on **soft robotics**, **morphogenetic design**, and **biofeedback-driven control systems**
- Early proposals on **quantum biology**, **field memory**, and **haptic AI architectures**

Final Note

Ripple theory is not a closed field. It's a tuning fork. These readings are **not prerequisites**, but **resonance points**—sources that harmonize, inspire, or challenge your ripple sense. Follow whichever one vibrates for you.

Because every ripple begins with one question that dares to listen differently.

Acknowledgments

For the silence, the friction, and the field.

This book would not exist without the strange, persistent sense that something in the way we build—how we think, how we move, how we land—wasn't quite aligned with the world itself.

To all the misfits of science, the rebels of design, the quiet engineers, the curious students, and the unlicensed dreamers—**thank you**. You kept listening when the equations stopped speaking. You sensed a rhythm long before we had a name for it.

I offer deep gratitude to:

- The **scientists and philosophers**—both historical and contemporary—who challenged what was known and revealed cracks worth crawling through.
- The **engineers who failed beautifully**, whose rockets blew up, whose wings snapped, whose probes drifted off-course, only to leave behind a trail of new questions that pointed toward something deeper than thrust.
- The **designers and artists** who worked without blueprints but always in flow, reminding us that beauty often arrives before understanding.
- The **students and skeptics** who dared to ask, "Why does this really work?" and "What if the field has a voice of its own?"

To every teacher—intentional or accidental—who helped me unlearn just enough to see motion not as force, but as consequence, I thank you.

To the **structured vacuum**—for being far more structured than the textbooks admitted.

And to the silence itself— For revealing, patiently and without drama, that **resonance is always an option**. That the universe is not asking us to push harder. It's waiting for us to **move in tune**.

May this book ripple forward—into hands that build, into minds that question, and into futures that finally learn to listen.

With gratitude,

Sandeep Chavan

About the Author

Sandeep Chavan is an independent researcher, educator, and author whose work challenges the deepest assumptions of modern science—by listening to what the field itself is trying to say.

With a background in engineering and a deep-rooted calling to reframe how we understand motion, force, energy, and intelligence, Sandeep is the creator of the **Universal Energy Dynamics (UED)** framework—a first-principles theory that redefines the very fabric of reality as structured ripple flows, not abstract spacetime or probabilistic particles.

His research in UED spans across disciplines—physics, cosmology, fluid mechanics, AI, consciousness, and design—and proposes bold alternatives to classical theories like General Relativity and Quantum Mechanics. Through dozens of published books and scientific reports, Sandeep has introduced key concepts such as:

- **Ripple Gradient (∇E)** as the true driver of motion
- **Structured Vacuum Energy (SVE)** as the active field behind reality
- **ATT (Alignment Threshold Time)** as a natural boundary for coherent motion
- **Ripple Intelligence Modules (RIMs)** and **Ripple-Conscious Systems (RCS)** for future AI design
- **Chavan's Limit** and **Chavan-Einstein Unified Theory (CEUT)** as foundational reboots for cosmological dynamics

His reports—distributed through platforms like Zenodo—offer mathematical formulations, thought experiments, and real-world test proposals for validating UED concepts. He has actively engaged academic circles and research institutions, urging the scientific community to re-express the universe through **consequence-based logic** rather than force-driven abstraction.

Beyond theory, Sandeep is a passionate educator who believes in **teaching under a tree if needed**—as long as it brings truth to light. He runs Gyrus Institute in India, where students are not just trained for exams but invited into inquiry.

Rocket Science is part of his broader mission to help humanity stop pushing against reality—and start moving with it.

For Sandeep, physics isn't a subject. It's a conversation between the mind and the universe. And UED is his way of making that conversation finally make sense.

Don't miss out!

Visit the website below and you can sign up to receive emails whenever Sandeep Chavan publishes a new book. There's no charge and no obligation.

https://books2read.com/r/B-A-EPGPC-YWGOG

Did you love *Rocket Science: The Ripple Gambling*? Then you should read *Chavan vs Einstein: The Universe Was Never Curved*[1] by Sandeep Chavan!

[2]

Chavan vs Einstein: The Universe Was Never Curved, *A Scientific Manifesto for the Next Reality*

What if gravity isn't a force?What if time isn't flowing—but delayed alignment?What if the universe doesn't need particles, forces, or curved space at all?

In this groundbreaking work, independent Indian Physicist **Sandeep Chavan** invites you to step beyond Einstein—not to dismiss him, but to finish what he began. *Chavan vs Einstein* introduces **Universal Energy Dynamics (UED)**, a revolutionary

1. https://books2read.com/u/bPjGqz

2. https://books2read.com/u/bPjGqz

ripple-based framework that redefines motion, force, time, gravity, and even perception itself.

Gone are the fix-it constructs of dark matter, dark energy, and singularities. In their place is a coherent, consequence-driven model where **Structured Vacuum Energy (SVE)** forms the field of reality, and every observed phenomenon emerges from **ripple tension and alignment thresholds**—not from abstract geometry or missing mass.

This book dares to say what others only question:

Space doesn't bend. It **ripples.**Gravity doesn't pull. It **invites**.You aren't just observing the universe. You're **a ripple within it**.

Through sharp logic, reframed thought experiments, and poetic clarity, Chavan dismantles the illusion of external forces and replaces them with a **field-first ontology**. Time becomes **Alignment Threshold Time (ATT)**. Motion is no longer displacement, but **realignment**. And consciousness? It's not a byproduct of the brain—but the **field remembering itself.**

Accessible to both physicists and curious thinkers, *Chavan vs Einstein* is part science, part philosophy, and part awakening. It bridges classical mechanics, quantum paradoxes, and metaphysical questions with a single, elegant language: **Ripple Logic**.

This is more than a book.It's a blueprint for the next scientific revolution.Not through more complexity—but through **radical clarity**.

Whether you're a scientist, philosopher, student, or silent observer—this work will shift how you see everything, from falling apples to fading stars. And perhaps most importantly, it will realign how you see **yourself.**

"The universe was never curved—we were.This is how we straighten."

Read more at https://www.sandeepjchavan.com/.

Also by Sandeep Chavan

Chavan's UED research

CMGED Research Report

Chavan's Cosmic Continuum (C^3)

Frequency

Ripples of Reality

Reality Realigned

The Great Realignment

I – The Cosmic Leverage

Chavan's Cosmic Theory of Relativity (CTR)

Chavan's Theory of Light

Feeling Like a Physicist

TIME – The New Definition

Echoes of War: The Russia-Ukraine Conflict and Its Global Ripples

Shattered Lives

The War Economy

Borders Redrawn

Arms and Allies

Collateral Damage

Justice in the Ashes

Somewhere in the Darkness of the Cosmos

The Blueprint of the Universe

The Cosmic Breath

From the Edge of the Universe

The Eternal Sage Chronicles

The Sage and the Sovereign

The Family Pillar

The People's Flame

The Warrior's Soul

The Bridging Mentor

The Guiding Light

The Exodus

The Last Cycle: 2075

The Fundamental Theory of Life & Existence

Energy in Action: The Soul Demystified

The Active Mind

Energy in Connection

Death: The Disconnection

Subconsciousness: The Chosen Disconnection

Desires: The Engine of Life

The Restoration of Human Knowledge

Restoring Bhagavad Gita

The War of Thoughts

Cracks in the Known

The Human Game

The Divine Dilemma

The Original Intent

The Answers

Ultra Modern Life Philosophy

The Duality of Laughter and Tears

Take It Easy

Compare to Accept

Standalone

It's Not AI, It's AHI - Amplified Human Intelligence

The Decision Paradox

The Four Sapiens

Malicious Script of Indian Polity

The IIT Legacy & Global Impact

The IIT Legacy & Global Impact

You are just a Version of Your Original

Win the Game You Didn't Choose

Always Keep Your Bags Packed

Reveal to Shield - Learn the Game of Facades

Engage Beyond Elections
Culture, Identity & Change: The Evolution of Indian Society
The Triangular Dynamics
Never Trust Without Doubt
If You Can't Change the Country, Leave the Country
Live Fearless with Fear
India: The Investment Magnet
Realignment
Powering a Billion Dreams
Universal Skill
Extension: From Brush to Algorithm
Human Side of Einstein
The Future Knows No Lines
Let's Bring the Generations to the Table
Unified Energy Dynamics
A Beat That Never Dies
Strategic Horizons
Right From Scratch
The Forgotten Innovators
Physics Fixing
The Savior Illusion
The Selfish Side
The Age of the Weakest Sapiens
Neural Wars
The Absorption: There Is No Death
Strange Therapist
Bubbles: Become a Drop
Beyond the Blue
Half-Scholar
The Law of Truth
Failure
Philosophy 2.0: A Structural Manifesto
Chaos: A Gentle Guide to the Unresolvable

Self-Help: New Syllabus
Rippling
Game: How to Play the Share Market
Out of Syllabus
Saffronized: The Code Behind the Crown
The Anger
Moksha: The Liberation
Chavan vs Einstein: The Universe Was Never Curved
Rocket Science: The Ripple Gambling

Watch for more at https://www.sandeepjchavan.com/.

www.ingramcontent.com/pod-product-compliance
Lightning Source LLC
LaVergne TN
LVHW090305210926
848480LV00001B/139
* 9 7 9 8 2 3 1 1 9 9 8 1 5 *